# FLYING
# IFR

*Also by*
**RICHARD L. COLLINS**

FLYING SAFELY

# FLYING IFR

by

# Richard L. Collins

*DELACORTE PRESS / ELEANOR FRIEDE*

Published by
Delacorte Press / Eleanor Friede
1 Dag Hammarskjold Plaza
New York, N.Y. 10017

Manufactured in the United States of America

First printing

*Designed by Oksana Kushnir*

LIBRARY OF CONGRESS CATALOGING IN PUBLICATION DATA
Collins, Richard L    1933–
    Flying IFR.
    Includes index.
    1. Instrument flying.   I.  Title.
TL711.B6C64      629.132'5      77-26917
ISBN 0-440-02651-2

*To Ann, Charlotte, Sarah, and Richard, Jr.,*
*who rode through a lot of those clouds with me*

# CONTENTS

# FOREWORD

Twenty-five years ago IFR flying in light airplanes was the exception to the rule. That has changed now, and on almost any IFR flight you can hear as many single-engine and light twin general aviation aircraft as airliners on the frequency. This is entirely as it should be, because flying IFR is what makes the airplane a reliable traveling machine. Where the VFR pilot is bedeviled by every cloud and visibility restriction, the IFR pilot's no-go situations are encountered very infrequently. With instrument flying capability we can really get our money's worth out of an airplane.

Flying IFR is not difficult, but there are a lot of things that can be learned only through experience. Discovery is a big part of the activity, because the many excellent manuals on the theory of instrument flying don't really cover the practical aspects, and often training is not conducted in actual conditions. The training concentrates on preparing the pilot for the FAA written and flight tests; then we are on our own, ready to explore and learn about the real world.

This book is about the real world. It makes no effort to prepare the reader for any tests other than the one of time. Regulations are covered only as they apply to practical situations and as they directly relate to IFR safety. No pet theory is offered on how to enter a holding pattern in the recommended (but not required) manner. The book is strictly about *flying IFR* in light airplanes, and it should be helpful to both the student of instrument flying and the practicing instrument pilot.

—RICHARD L. COLLINS

# 1

# THE FOUNDATION

One of the primary tasks in instrument flying is setting priorities straight in the mind. I found a good illustration of this one day as I followed another airplane on an ILS approach in VFR conditions. The pilot ahead was having a terrible time, swooping, dipping, and making what appeared to be rather mad lunges at the ILS course with his airplane.

Clearly this pilot had no business practicing ILS approaches. It was purely wasteful—he had the cart before the horse. From two miles behind it was plain that the pilot didn't have a strong capability at basic instrument flying. He could not assimilate the data on the panel, he could not hold a heading, and he could not maintain a constant rate of descent. These problems are rather common; in this modern day it is easy to forget the basics as we rush headlong into flight that seems related more to vast arrays of talented transistors and computers than to one person and one machine.

Regardless of a plane's sophistication, the basic ability to fly the airplane is first in importance. If you can't fly the airplane solely by reference to instruments, with a measure of precision and confidence, there is no way to utilize all these transistors and computers. Basic instrument flying is as important today as it was to Charles Lindbergh in 1927. It is top priority.

Some years back, when the Aztec first came out, I had one around to fly for a few days. The first order of business was some simulated instrument flying under the hood, and the first thing I did was shoot a VOR approach. The airplane didn't seem to want to come down quickly enough, so I put some flaps out when passing the station on the way to the airport. That was okay. On the missed approach, though, everything went wrong at the same time. I retracted the gear and then put the flap selector in the Up position. The nose on the first Aztecs becomes very heavy when the flaps are retracted, this is backward to what you usually find in a low-wing airplane, and it wasn't what I expected. Between reaching for the trim, pulling on the wheel, and wondering what in the devil was going on, I neglected the basic art of flying for a moment. The airplane began to go its own way. The safety pilot suggested I look up and fly visually to get things under control. It was embarrassing.

## LESSONS

The incident bore more than one lesson. There was a clear message about learning something about a particular airplane before operating it on instruments. Delve into *basics* about trim changes with gear and flap retraction and extension, learn power settings for various phases of flight, and study the machine and its operating handbook for things that are different or less than obvious.

I certainly should not have chosen a simulated instrument approach as the first chore in the airplane when flying it under the hood. Instead, I should have flown it around for a bit, developing a relationship between my mind, my hand, and the airplane's characteristics. Every pilot/airplane relationship starts out with the pilot a stranger to the machine, almost in the role of a passenger in the left front seat, and evolves to the point where the human almost becomes a part of the machine. With the Aztec, I skipped the introduction and instead tried to immediately develop a complex relationship.

The most important lesson is to heed the message from any such evidence of a problem with basic instrument flying. Be a perfectionist about it. If you can't do an excellent job under the hood, the situation surely is not going to get any better in cloud. A license to fly instruments is proof only that the pilot was capable on the day the license was obtained. After that, it is up to the pilot to stay current. When some weakness shows in practice (or actual) instrument flying, don't pass it off with the thought that it couldn't have been too bad "because I made it." If the flying isn't something to be proud of, take it as a mandate to practice and polish basic instrument flying.

---------------- **TWO WAYS** ----------------

I think that we fly an airplane by reference to instruments in two distinctly separate ways. One way is natural, the other is purely mechanical. It is important to identify and use both methods to maximum advantage.

You have probably observed what might be called a "natural pilot" at one time or another. Flight is seemingly without effort. When VFR, little time is spent looking at the panel, yet the altimeter stays glued, and the omni needle dead-centered. It's the same IFR.

This is done with a keen awareness of the attitude of the airplane. In smooth air, if the nose is kept "right there" and the power is correct, the altitude of the airplane will not change. If the wings are kept level, with the ball in the center, the heading will not change.

It is a relaxed way to fly, a result of the ability to perceive and respond without conscious, mechanical mental effort. In VFR conditions, the view out front and to the sides combines with the sound and feel of the airplane to tell the pilot what is happening. In IFR conditions, the expanse of the view outside is compressed into the artificial horizon. The information is there, and the sound and the readings of all the other instruments in the airplane verify that things are going well. The instrument cross-check, or scan, is talked of with reverence, but the natural pilot might not think in terms of a scan. The ability is in absorbing the big picture, in using all the human senses to fly the airplane.

―――――――――― *ALERT* ――――――――――

This isn't to say that the natural pilot does not scan the panel. If asked for the oil pressure, the response might come in an instant. If a generator or vacuum pump failed, the pilot would catch it, through the instrument indication, in seconds. The pilot is simply able to operate with a single-mindedness of purpose, thinking of nothing but flying the airplane, and to gather all the data on the instrument panel in a relaxed and informal manner.

―――――――――― *MECHANICAL* ――――――――――

The instrument pilot flying mechanically might be the same human as the one who was just flying naturally, but on a different flight or a different portion of the same flight. There simply comes a time when it is necessary to fly with absolute procedural discipline. In the beginning of that early Aztec flight, I was probably flying the airplane rather naturally. I was moving along well and was seeing everything. I was relaxed, loose as a goose. Things were fine. Then, when the flaps were retracted, my mind jumped track. Instead of seeing everything, I saw nothing. I only wondered about the unexpected trim change. I did not make the required transition from natural to mechanical flying when a problem made this necessary. At the first sign of something unusual, the clear call was for absolute attention to basic instrument flying. I should have disciplined myself to base eye and thought on the artificial horizon, and to put the airplane in a wings-level climb attitude. Then a mechanical and methodical check of the other instruments should have been activated. Airspeed on the proper value. Altimeter indicating climb. Heading steady. Turn needle (or turn coordinator) steady and straight. Vertical speed indicating rate of climb. Discipline. Once that mechanical process was in place and working, I could have expanded to include missed approach procedures, and navigational chores, and perhaps I could have allowed myself three seconds to ponder the trim change with flap retraction.

Some pilots never feel they have reached the point where they are flying instruments in a natural manner. There's nothing wrong with that. The mechanical way works just as well, and remember, the person who does it naturally much of the time still has to revert to mechanical means at certain times. And if you miss the cue, things are bound to get worse before they get better.

## FROM THE START

The beginning of an instrument flight is when almost all pilots fly mechanically and when the ability to concentrate on the basics is extremely important. An instrument pilot who flies infrequently might get the first dose of actual IFR as a climbing and accelerating airplane punches into the bottom of bumpy clouds. The key here is in concentrating only on flying the airplane until comfortable with it.

—————— *ACCELERATION ERROR* ——————

There is a special consideration in the initial climb that can be used to outline a couple of technique items. Often the artificial horizon does not appear quite normal in the first phase of a climb. The depiction of nose-up attitude is accentuated by an acceleration error that makes the nose-up attitude appear more pronounced than it really is. The nose-up attitude in a Cherokee might for a few moments look more like what you would expect in a jet. A pilot might misinterpret this, lower the nose, and fly back into the ground *if* the indications of other instruments are not included in the deliberations.

First technique item: A pilot can and should watch instruments during noncritical times, VFR, and relate the picture on the panel to the view out the windshield. As a result, the acceleration error will be a known quantity and the pilot will know the indication when the pitch attitude is correct. That is what instrument flying is all about—knowing what it would look like if all the clouds went away.

Second technique item: It is in the climb that we first put to the test the ability to look at the correct things at the proper time, get the message, and make the necessary control movements. What do we look at, and how do we demand proper performance?

To begin, base on the artificial horizon, just as you would fly the correct attitude by referring to the real horizon in VFR conditions. Then grade the attitude being flown with a scan of the other instruments. When you are flying a standard instrument arrangement, a glance to the left will reveal that the airspeed is steady on the proper value; a glance to the right will show that the altimeter is moving upward. A glance down shows that the heading is steady on runway heading. The turn-and-bank or turn coordinator will verify that the airplane is indeed not turning, and the vertical speed is double verification that it is climbing.

## ATTITUDE

The emphasis is on attitude, on the artificial horizon. It is home base for the eyes, with the scan of the other instruments verifying that the selected attitude is producing the desired results. In the case of initial climb, the desired result is climbing while flying at a predetermined airspeed. But we still use the artificial horizon as a primary reference, because it tells us about bank and pitch attitude simultaneously. No other instrument on the panel does that trick.

## SCAN

In these first moments of instrument flight we must come to grips with some method of scanning the instrument panel. Some call it a "cross-check," a term that came from the military.

When I worked at an Air Force contract school, much emphasis was put on cross-checking. The emphasis was good, too, but nobody ever explored how it is done. The military did teach in terms of primary and supporting instruments, to put emphasis on the most important things at various times of flight, but there was no explanation of how much time should be spent with each instrument. In fact, the Air Force manual stated: "It has long been known that pilots do not use any specific method of cross-checking, but that they do use the instruments which give the best information for controlling the aircraft in any given maneuver. Most of the pilot's attention is devoted to checking these important instruments." That's rather like saying that most pilots don't plan a takeoff run longer than the available runway.

The FAA is no more definite in saying, in reference to scanning: "There are no set rules and no single method." All of which leaves the reader wondering what the devil to look at and how much time to spend on it.

———————————— *EXPERIENCE* ————————————

My experience has been that when flying instruments in smooth air I basically look at one thing, the most important thing (often it is the artificial horizon), and check the rest with peripheral vision and furtive glances. In a climb I'm likely to look solely at the artificial horizon and satisfy the requirement of scanning with peripheral vision. The airspeed needle can be seen out of the left corner of the eye, on a correct value; the altimeter reading out of the right corner is okay if it is increasing. I can see that the vertical speed is on a positive value and the turn coordinator is visible to the lower left. Right at first I don't worry that much about a precise heading—if the wings are level, it will remain close. The primary thing in my mind is establishing myself at the chore of flying instruments. The task is controlling the attitude of the airplane. This is done by looking at the artificial horizon. If a rate instrument suggests the need for a change in attitude, that change is made while I am looking at the artificial horizon. In short, I don't try to control the rate instruments directly.

## SECOND THINGS SECOND

The intial IFR clearance probably includes instructions to turn to a heading after takeoff and to contact departure control. The pilot must take first things first, though, and use the first minute of the climb in instrument conditions to make friends with the airplane, to settle in with the task, and to defeat any onset of spatial disorientation. If this isn't done, success is impossible. Once it is done, call departure control and then turn to the assigned heading.

As we progress through a flight, flying becomes easier. The airplane is more familiar, and if the flight remains in cloud for an extended period of time, the clouds and the water streaking back along the windows become friendlier. It is at this time that almost every instrument pilot lapses into more natural flying. Ease the seat back a notch and absorb all the messages from the panel. You can learn as you fly along, too, by monitoring what you look at and, if the results are good, storing this for future reference.

For example, you'll note that peripheral vision works well so long as the air is smooth but becomes more difficult to use when passing through turbulence. You just don't get a clear message out of the corner of an eye when the airplane is jiggling around. The call is for more of those glances to the other instruments, to verify that the attitude selected on the artificial horizon is doing the job.

## CHANGES IN THE PATTERN

You'll notice changes in pattern, too. For example, during a descent in smooth air I noted that I was looking primarily at the directional gyro. The controller had assigned a heading, and that heading was occupying my attention. So long as it remained steady, I knew the wings were level. Peripheral vision verified this with the artificial horizon. I was descending to an altitude, and an occasional glance at the altimeter gave the progress on that. The airspeed was easy to interpret with peripheral vision, because the needle was close to the top of the green sector on the indicator.

## HAND-FLY

You can't learn much about your basic instrument flying with the autopilot doing the work, so hand-fly the airplane when in cloud. It is a fine time to practice precision instrument flying. Hold the heading and altitude precisely. Keep the navigation needles centered. Work at the division of time between instrument flying and chores such as frequency changes, calculating estimates, consulting charts, and writing down revised clearances. Fly it for thirty minutes or an hour perfectly and then turn the autopilot on if you wish—but earn the rest first.

## BASIC MANEUVERS

The elements of basic flying tasks are things that can and should be practiced. There is really nothing to instrument flying other than climbs, climbing turns, level flight, level turns, descents, and descending turns. Those are the things to work on until they are well in hand. The rest is pointless until you can fly with precision. Grade every flight, and work at the basics methodically.

## TOUCH

Control touch is an important part of instrument flying. For an illustration, trim the airplane, release the controls, and then don't touch except as necessary to correct an instrument reading. Touch with only one finger. Nothing is likely to get far off, and you'll soon see that you could fly all the way across the country using just one finger. The airplane requires very little "flying"; it is more a matter of a pound of pressure here and a pound of pressure there—at the correct time. Contrast this with the oft-noted jut of a pilot's jaw when heading into instrument conditions. Some look as if it is to be an instrument fight instead of an instrument flight. Back to our pilot who was having trouble with the ILS: From another airplane it was obvious that he was making abrupt and gross corrections, that he was fighting with his airplane.

## FIXATION

Manual and visual fixation is acknowledged as a scuttler of instrument pilots. Building a guard against this is part of the basic art. It is often tempting to hang on one instrument, to stare. If the airplane is a hundred feet low, a pilot might look at the altimeter, add a bit of back pressure, and wait for the altitude to come up a hundred feet. All the while, the bank attitude of the airplane might be going to pot. The correct way to fly is to note the excursions and trends of instruments and then use the artificial horizon to change the attitude of the airplane in a manner that will nudge any wayward instruments toward a proper value. Don't move a control without consulting the artificial horizon, and keep an eye on the horizon during the control input. Then check results on the other instruments. This keeps the eye and mind active. It does the job.

Fixation can take many forms. We can, for example, lapse

into daydreaming: The eyes are on the instruments but the brain is on the note at the bank. Or, in time of trouble, the mind might be virtually paralyzed by a problem such as turbulence or mechanical malfunction.

I've found the best cure for fixation to be a thorough tongue-lashing. I speak to myself frequently at such times, rather sternly. The admonition is usually to settle down and fly the airplane. The reward offered is that all bad things will pass if the airplane is flown properly.

## ─────── POWER VERSUS ELEVATOR ───────

Power is a flight control that we use in basic instrument flying. The use of power is both very important and very simple. In a basic airplane you really need only five settings, but you need to memorize them so that power can be quickly set for what you want to do; then attention can go to the proper pitch attitude. The basic power settings are: climb, normal cruise, cruise in turbulence (maneuvering speed), normal descent, and descent in turbulence. In faster airplanes and retractables, a few more settings are needed for instrument approaches. They would be for level maneuvering or holding at reduced speed, and for final approach descents with the wheels down in the case of the retractable.

One further thing must be considered in relation to power. There has long been an argument about whether power controls airspeed or altitude. The same argument is applied to the elevator control. Hopefully the instrument pilot is too savvy to fall victim to an argument on this score. In certain situations it is best to think of power as a primary influence on altitude and in other situations it is best to think of the elevator as a primary influence on altitude. For example, if you are flying level at 120 knots and the time comes to start down, you don't want to think of the elevator as the control to use. Power would be the thing then. Reduce the power to begin the descent and maintain the airspeed with the elevator. By the same token, if you're running just a tad high on the glideslope, lowering the nose a hair would be a perfectly acceptable way to make that altitude correction. Or you could back off the power a bit. Anyone who would argue with either would be nit-picking.

The business about what controls what is critical only when the airplane is being flown near some extreme. Extremes are unnecessary in light-airplane instrument flying and should be avoided. Just for the record, though, remember that in low-speed situations where the chips are down, using the elevator to control airspeed is what will save your tail.

## SELF-TAUGHT

There's one recurring thought about the basics of instrument flying: They are not things that anyone can teach a pilot, they are things the pilot must learn through experimentation and experience. No instructor can tell exactly where you are looking, and no instructor can know when you are experiencing a touch of the "leans" or when there is fixation on some instrument or subject. The pilot must practice enough to work these things out, and to determine where the eyes need to look at given times. What an instructor can do is give you helpful hints, such as recommending 15 inches of manifold pressure here and 13 inches there, but even that must be subjected to a measure of self-discovery. The instructor can also critique your basic instrument flying, but if you'll be objective you can do an even better job there, because only you know where the mind and eyes were when a mistake was made. The instructor can pinpoint the result; only the pilot knows the cause.

# BASIC VARIATIONS

Now that we are proficient and current on those basics, can we spend all our hood time on those exciting (and required) instrument approaches? Huh, Teach, can we? Please?

Patience, friend, because in instrument flying patience is the only virtue and impatience is the only sin. There are variations on the basics, and mastery of these can be a real life-preserver in time of need.

Partial-panel is the primary variation. This can encompass the loss of one or more instruments. In single-engine airplanes it is highly pertinent, because there's no redundancy of power sources and failure of either the electrical or vacuum power source will disable some of the instruments.

Vacuum-source failure is a problem that is often addressed. This results in almost immediate loss of the artificial horizon and the directional gyro in most airplanes. The hori-

zon is the cornerstone of our full-panel flying, so this is quite
a blow. And the DG (when properly set) is the primary heading indication, so its loss is also quite serious.

Be optimistic, though. At least there are fewer instruments
to demand attention in this situation. Only the turn-and-
bank (or turn coordinator), airspeed, altimeter, and vertical
speed require attention. And the basic requirements of flight
remain the same.

To begin, make friends with the instrument that gives information that can be related to the bank attitude of the
airplane. The airplane can be trimmed to a stable longitudinal situation, but lateral stability absolutely must be handled by the pilot.

Either the turn-and-bank, with a needle, or the turn coordinator tells the tale. For general discussion, both will be referred to as a turn indicator, and think of either as simply
telling whether the airplane is or is not turning, the direction
of any turn, and the rate of turn. There is no direct bank-attitude information there, but the bank attitude can be envisioned with the information from a turn indicator. Fly with
a mental picture of bank, and use bank (ailerons) to control
the instrument.

Three primary bank-attitude tasks must be mastered. A
straight-ahead condition comes first. If the turn needle is kept
centered, or if the turn coordinator is kept level, the heading
will not change. As noted earlier, being able to hold a heading is a most important basic. If the ball is in the center, the
airplane will also be in level coordinated flight, but don't
worry a lot about the ball. If you hold the heading, who cares
if the ball is a tad off one way or the other? If it is off more
than a tad, do something about it with the rudder or rudder
trim, applied in the direction of ball deflection.

Next comes a turn in each direction, first at a standard rate and then at half a standard rate. Three degrees per second is the standard, and both types of turn indicator have an index for this rate of turn. The rate of turn is important if the directional gyro is gone, because the magnetic compass is difficult to use in turning to headings. So we use the beginning heading and combine it with the rate of turn and time to tell us when we have reached a new heading.

## THE VISION

How much bank do we see in a standard-rate or half-standard-rate turn? If you are looking at the needle of a turn-and-bank, nothing gives a mental image; the pilot must interpret. The bank required for a standard-rate turn is determined by airspeed, and you should know a couple of values for your airplane: the angle of bank for a standard-rate turn at normal cruise and the angle of bank for a standard-rate turn at a reduced maneuvering speed. Roughly, the airspeed less the last zero and plus seven is what it takes. For example, at 100 knots a standard-rate turn requires about 17 degrees of bank; at 140 knots it would be 21 degrees. Half standard rates would be half that.

The turn coordinator shows turn by the banking of a little airplane, and the variation in bank angle with speed is one reason I do not like this instrument as well as a turn-and-bank. Remember, it shows rate of turn, not bank, even though the presentation purports to do the latter. Going to an extreme for illustration, it shows an indication of what appears to be the same angle of bank when an airplane is actually banked 15 degrees at 80 knots or banked almost 30 degrees at 200 knots. This can be misleading if a pilot hasn't put some effort into understanding the shortcomings of this instrument.

--- **EXPERIMENT** ---

As you experiment with partial-panel, you'll probably note that the turn indicator requires the most attention, and if you give it that attention the airplane is rather easy to fly. If the airplane is properly trimmed, pitch control is going to involve only a pound of pressure here and a pound of pressure there as the airplane is maneuvered. The real task is managing bank attitude. When pilots become spatially disoriented and lose control of an airplane, bank is what they have lost control of. Keep the wings of a properly trimmed airplane level and nothing much happens. Bank the airplane, or let it bank unintentionally, and everything starts to go awry.

Try a little experiment next time you fly. With the airplane trimmed and the wings level, as shown by a centered turn indicator, note that things are stable longitudinally. Then bank until a standard-rate turn is indicated and release the controls for a moment. Only bank attitude was changed, but it affects a lot of other indications. The airspeed, vertical speed, and altimeter will all make a move very quickly.

Even though there are fewer instruments to look at, the instrument scan is more difficult on partial-panel than full.

With a standard instrument-panel arrangement, the inoperative instruments are in the center, leaving a large dead space in the middle of the usable panel. Too, the inoperative instruments might likely present misleading or confusing clues. The artificial horizon is likely to park in a bank when it runs down after the failure of vacuum source, and the DG will rest on one heading after spinning around. If peripheral vision catches the erroneous message from one of these and it registers on the subconscious, it might prompt at least the beginning of a misguided control input. It is good to have some means of obscuring the inoperative instruments from view in this situation. You can use the same covers used when practicing partial-panel. Keep them in your flight kit.

## PITCH

There are three good pitch-attitude references in partial-panel, and any one of the three can be used to visualize pitch attitude. They are most commonly used together, with one assuming somewhat more importance than the others during various phases of flight. For example, in level flight the altimeter tells of a constant altitude or of the need for a correction. If a hundred feet needs to be lost, for example, lower the nose slightly, note descent on the vertical speed, and as the altitude approaches the correct value level off. There is lag in the standard vertical-speed indicator but if all corrections are gradual, the instrument is quite useful. A little practice will teach its lag properties.

## HELPER

There is another useful instrument on most IFR panels. The ADF can be used as a stable indication of heading by tuning it to a strong station some distance away from the aircraft. Then, so long as the ADF indication is steady, the airplane is not turning.

## LIMITS

Very definite limits should be set on what will be attempted when flying on partial-panel in an actual situation. The standard-rate turn should be the absolute most bank that will be allowed. Rapid descents and climbs should be avoided. (Climbs should be almost unnecessary, as the failure of a vacuum system would be a mandate to go somewhere and land for repairs. Thus, a descent would be the most likely requirement.) In an actual situation partial-panel should be treated as an emergency, with the air traffic controller notified so that he or she can render as much assistance as possible and cut some corners for you if necessary.

## ADDED ATTRACTION

Fly straight and level now, with the eye perched mainly on the turn indicator and cross-checks on the airspeed altimeter and vertical speed. Next, we'll add a task. Bring the nav instrument into the picture. It is reasonably difficult to get somewhere in the clouds without navigating, and the VOR needle can be used in some basic partial-panel flying even though it is not directly sensitive to heading.

What you'll learn rather quickly is that, given some talent with the turn indicator, it is surprisingly easy to fly to, and right over, a VOR station without continually referring to the compass. This is based on being able to hold a heading, and those who are both proficient and motivated often do as good a job of holding a precise heading with the turn indicator as they do with the DG. When flying with the horizon and DG it's common to not even look at the turn indicator. Who needs it? It's also common to weave a little bit—a few degrees to the left and a few degrees to the right. In partial-panel flying, turn indication gets so much attention that the pilot tends to *not* let the heading stray much. For instance, if the airplane banks slightly and the turn indicator deflects half a standard rate indication to the right for one second before it is caught, the heading will have changed only one and a half degrees. You can hardly see a degree and a half on a DG.

---

## *TRACK*

---

In tracking to a VOR station, it is true that the compass must be referred to quite often when you are a considerable distance from the station. The VOR isn't heading-sensitive— it only gives information about the airplane's position relative to the station and desired track—so if you are, say, thirty miles from a station, the needle really won't let you know that your heading is incorrect very quickly, and if you don't know what heading resulted in the deviation it is difficult to decide on a correction. As the airplane gets closer to the station, though, tracking can be done very accurately with just the turn indicator and the VOR needle.

How?

The key is to be able to hold a constant heading with the turn indicator, and to be able to make accurate 5-degree heading changes. These small turns are best made at half standard rate, with the turn held for a count. Practice with the DG uncovered to make perfect.

With the airplane stabilized on an approximate inbound heading—for example, 180 degrees—and with the VOR showing 180 "To" with the navigation needle centered, maintain a constant heading with the turn indicator and keep an eye on the needle so that any movement will be promptly noted. If the needle starts to drift to the left, for example, make a mini-turn (5 degrees) to the left. Then nail that turn indicator back to the center and watch the needle. If it persists in a drift to the left, make another mini-turn to the left. If it remains steady, good, unless it's more than half-scale out, in which case another turn to the left would be dictated, to start the needle back toward center. Then, as it approaches center the correction could be taken off. Success is found in being on an approximately correct heading to begin with, so that any needle movement won't be drastic, in making all corrections small turns of a known value, and in holding a heading between corrections. As the station gets closer, excursions of the VOR needle will be at a more rapid rate. And if it moves out to one side and stays there, the limit would be a couple or three little turns toward the needle and then a steady heading to await a "To" to "From" switch at station passage. After you try this under the hood a few times, you'll find that the turn indicator is very trusty in tracking with the VOR.

<hr>

## THE ADF AGAIN

The localizer works the same way, except here there's a wild card if you have an ADF. As noted, that ADF can be an indicator of a steady (or changing) heading. It also gives positive relation to a point on the localizer course, the outer compass locater.

Any time the localizer needle is in the center and the ADF (tuned to the outer locater) needle is on zero, the aircraft heading is the same as the localizer and the ADF is pointing directly through the nose of the airplane at another point on the localizer. When off the localizer, the ADF provides orientation to a point on the localizer.

An example of how useful the ADF can be is in intercepting the final approach course. Position the aircraft outside the outer marker, either just completing a procedure turn or in the process of intercepting the localizer for a straight-in. The localizer being intercepted runs northeast and the airplane is south of the localizer, headed approximately north.

So long as the ADF needle is kept to the right of zero and the localizer needle is to the left, we know we will intercept the localizer outside the outer marker. It is like flying into a funnel. Now, what happens if the ADF is actually used for heading information and the needle is kept 45 degrees to the right of the nose? Simple. We intercept the localizer outside the outer marker. When the localizer needle starts to center, turn right until the ADF needle is on zero. Presto: The airplane is on or at least near center line of the localizer inbound, and the heading is approximately the same as the inbound localizer heading. From here the localizer is flown just as the VOR was flown when close to the station—using mini-turns to correct for drift and the turn indicator to hold a heading precisely between adjustments.

The ADF can be used like this approaching the localizer from either side; the headings and position at the beginning of the discussion were just to give you a mental picture of an approach. Practice using the ADF in this manner, because it is an important tool in basic partial-panel instrument flying.

## GLIDESLOPE

Flying glideslope on partial-panel is a worthy subject in a discussion of basics, because it can seem a difficult thing to do at a crucial time. The concentration required to track the localizer just doesn't leave an abundance of eye and mind time for tracking something else.

A patch on normal procedures can thus be in order. To insure getting down after passing the outer marker, descend below the glideslope to the point where the no-glideslope minimum descent altitude is reached, then level off at that altitude. When the glideslope is intercepted again, from below, a few more feet can come off; you can even go on down to the full ILS decision height, depending on how good the glideslope interception and tracking proves to be. If the airplane doesn't have a glideslope, you can forget that problem. The key thing is that the localizer needle is of vital importance all the way in. If it ever drifts away to a full-scale deflection, the approach must be abandoned. So long as you don't go below the no-glideslope minimum descent altitude, the glideslope isn't quite so important. What glideslope does in most cases is lower the minimum from 300 feet and ¾ mile (or 4,000 RVR, if you will, and 300 MDA) to 200 and ½. And we trust that the day your vacuum system or gyros flunk out you will have the good fortune to find a place with a little better weather than 200 and ½ for your grand arrival.

Practicing all this can be quite challenging. First work at the basics, and then apply the basics to getting down. You might say to yourself: "The localizer intercept is strictly no sweat using the ADF on the outer marker. I am going to try to fly the glideslope all the way in. When the marker is reached inbound, the localizer is reasonably well tied down. Put the landing gear down to start the airplane on a rate of descent that will approximate the glideslope. The communication with the tower passing the marker cost a localizer excursion. A couple of mini-turns got that in shape, but it should not have happened. Learn to communicate *and* fly at the same time. If I can't, let the communication wait.

"The pressure builds on something like this, even though it is just under the hood. Maybe that is because it would be embarrassing to completely botch it in front of a trusting safety pilot. On this first one the control movements are a little stiff. Calling the tower at the river (when the safety pilot says we are there) results in another excursion, this time of both the localizer and the glideslope. The localizer is tied back down before the middle marker, but the glideslope needle stays down, showing the airplane to be high. High is okay. When the marker beeps through the sweaty atmosphere in the cabin, the altimeter shows 300 feet above the touchdown zone instead of the proper 200. A missed approach is done from this point, without looking up.

## GOING AROUND

"The sensations on the first partial-panel go-around in a long time, or the first ever, are pretty spectacular. There is no trusty artificial-horizon information about attitude. It is still quite flyable, though. The airspeed was on 100 knots when the go-around was started, so it can be primary reference for pitch attitude with no change in value. Apply full power and hold airspeed constant. As soon as the vertical speed shows a positive climb indication, retract the landing gear. The trim change is momentarily disturbing. The turn indicator has been approximately in the center all this time, but the localizer is long gone, full scale.

"The next approach is a little better, with less sweat. Don't really fly the glideslope. Rather, gradually descend below the glideslope to the no-glideslope minimum descent altitude and level there. Then, if everything else is just right, descend on the glideslope to the ILS decision height after intercepting the glideslope from below."

If such a practice flight doesn't reveal plenty of skill, go back to the practice area and polish on the basics in need. I find the most difficult time to be toward the end of the approach, when needles are moving faster and the temptation is to make larger corrections. I practice to cure that by working with simultaneous half-standard-rate turns and changes in descent rate of about 200 feet per minute. The objective is to do them smoothly and consider these values as maximums. Another good partial-panel practice is a simultaneous standard-rate turn and 500-foot-per-minute rate of descent that reverses at the end of one minute and then continues for one more minute. If done perfectly, the airplane will descend 500 feet and turn 180 degrees, then climb 500 feet while turning 180 degrees in the opposite direction. Exactly two minutes after the start, the airplane will be right back where it started. There is more discipline in that than you might imagine, and it will provide quite an education in what to look at to juggle two balls at once. Don't practice it just in smooth air—try it in the bumps.

## HOW TO TELL

A very fair question that is related to this subject involves the transition from full- to partial-panel. When the vacuum pump (or one of the instruments) shoots craps in actual conditions, how does a pilot man the bastions, catch the failure, and move from flying with the most to flying with the least?

To begin, it sure helps to know the basics of your airplane. If the instrument scan is active enough to catch a failure *before* it affects anything, and you know exactly which items are about to stop working because the suction gauge (a pressure gauge on some airplanes) went to zero, some preparations can be made. Such a malfunction will usually involve the artificial horizon and directional gyro plus the autopilot. (Ironically, the simplest wing-leveler autopilots are the best in time of vacuum failure, because most get data from the electric turn-and-bank or turn coordinator, and are electrically operated.)

Some airplanes have a vacuum-failure light. On these the odds are very much in favor of the pilot's catching a failure before it affects the instruments, because it takes a while for the gyros to spin down. If the airplane does not have a light, a pilot might miss a little gauge needle's movement to zero. The gyro instruments will tell you soon enough, though, and if the scan is active, the dying instruments shouldn't cause any control problem. My experience has been that the artificial horizon gives the first indication, with its pitch indication affected first. The airspeed, altimeter, and vertical speed will all argue with an erroneous pitch indication, and the alert pilot should easily survive this. (One thing that helps is to watch the gyros run down after the engine is shut down at the completion of a flight. Note how long it takes and what the instruments do as they park themselves.)

The mind is probably stimulated at the time of an actual vacuum failure, and the immediate transition away from the gyros isn't that difficult. My problem comes about three or four minutes later when the mind and eye keep trying to revert back to the old and complete way of flying. I've developed a pretty fair case of the uncomfortable squirms at such a time, and find it extremely helpful to use covers for the inoperative instruments.

It is very easy to lapse into a feeling of security when flying instruments, and to rely on the full panel and the autopilot in every operation. That's putting a lot of eggs in a complicated basket, though, and a pilot who does not stay proficient at the partial-panel variation of basic instrument flying can expect to encounter some very bad moments if the situation should arise. I would add that I've averaged about one vacuum-pump failure for every 1,500 hours of flying, so it is not a remote possibility. A problem in the electrical system is likely to happen more often, but an electrical problem usually leaves the pilot with a charged battery to parlay into a view of a runway somewhere, and will be discussed in Chapter 11.

# BASICS PLUS

Whether flying with all the instruments on the panel working or only some of them, the basics of instrument flying must be adeptly and methodically applied to all situations for the result to be a smooth and safe flight. There's no way to do this on a helter-skelter basis. All aspects of the things that we do in instrument flying must be examined; the result of the examination is the best way to do a particular thing.

For example, the pilot's knowledge of power settings for various phases of flight is acknowledged as important, but the application of that basic is not always simple. What if the pilot is descending at a speed in excess of the landing-gear extension and/or approach-flaps speed, is still a thousand feet above the glideslope-interception altitude, and the controller calls with word that the airplane is five miles from the marker and cleared for the ILS approach? If all our practice and search for understanding has been on the basis of decelerating to the gear speed while flying level, we are likely to experience some confusion when faced with a combination of descending and a need to slow down.

The message is to look carefully at all situations in practice. If your airplane takes four miles to slow from a normal letdown speed to landing-gear-extension speed in level flight with the power set on 15 inches and 2,200 rpm, you would know that in the example given there would be no way to arrive at the outer marker at the proper speed while following the accepted procedure. An unprepared pilot would have only blank thoughts; a prepared pilot would know there exists a choice between asking the controller for a 360 (which, in a busy area, can really clog the works) and pulling the power back below 15 inches, accepting the increase in wear that results from rapid engine cooling. And if the decision is to throttle back, that should have been done at least once in practice so you could note the trim changes, sensations, and sounds associated with the maneuver. The purpose of practice is to do all the things that might be expected in actual operations. So look for situations like this and work on them. Don't practice just the easy and routine things.

## TURBULENCE

Turbulence is another area where the application of basics is very important. The bumps can shatter calm and destroy a pilot's ability to perform unless there is confidence in a strong ability to manage the basics—wings level and a reasonable pitch attitude—regardless of the level of turbulence.

The best way to discuss instrument flying in turbulence is in relation to severe turbulence—a thunderstorm, if you will. From there you can work yourself back to anything else.

## NO WARRANTY

First, be assured that this comes with no money-back guarantee. We fly airplanes which are designed mainly around a 30-foot-per-second vertical-gust tolerance. (Some recently certificated airplanes have a somewhat greater gust tolerance.) Thunderstorms have soft spots and hard spots. As a storm develops, a tremendous vertical surge of air can develop at rates of ascent much greater than 30 fps. This happens in a limited area, and it doesn't last long because the energy is used up pretty quickly in such a squirt. It's there in some storms some of the time, though, which should be enough to make us try to stay away from all storms all the time. By so doing the encounters should be kept to minimum and the chances of flying into something that would mean automatic disaster should be remote.

The mechanics of thunderstorm turbulence will be discussed in Chapter 8. The question here is how to best fly a light airplane in a thunderstorm in the unhappy event that a pilot should make a long string of mistakes and wind up in the teeth of one. Again, no guarantees, but most areas of most storms are quite survivable, given the proper flying techniques.

The way to go is to fly almost entirely by reference to the artificial horizon. The goal is to maintain lateral control and a predetermined pitch attitude that will yield maneuvering speed, the best speed in turbulence. The power should initially be set at whatever yields maneuvering speed. It is a setting you should know in advance, and it is best to establish the airplane at maneuvering speed well in advance of any turbulence penetration. The seat belt should be *very* tight. The less you bounce around in the seat, the easier it is to see the instruments.

---

## THE UPDRAFT

The first thing you'll likely notice when flying into a storm is action associated with the air rushing into the storm and feeding the updraft. This will often tend to make the airspeed increase as the area of turbulence is reached, and if it is already high this is going to make any last-minute slowing to maneuvering speed quite tedious. It will also subject the airframe to potentially damaging stresses until the airplane is slowed. Those who hit an area of turbulence while going fast and trying to descend have problems on top of problems.

Once in turbulence, power can be used rather grossly in light airplanes with some degree of success. If the updraft is strong and is moving the aircraft upward at a great rate *and* causing an airspeed increase above maneuvering speed, don't be bashful about pulling the throttle back. Just maintain a level pitch attitude and pull the throttle on back. Certainly you want to do something about a large airspeed excursion, and power is the logical control to use here. Pulling the nose up would both exaggerate the G-loading from any bump encountered and maximize the altitude gain from the updraft. Pulling power back would tend to help contain the airspeed without bad side effects. There are a lot of stories floating around told by pilots who gained thousands of feet in thunderstorms with the power back to idle; that the stories are told must be some indication of success.

The emphasis must be on keeping the wings level. Turning is something to avoid in a situation where control is in doubt; we will explore the weather-related pros and cons of turning around in thunderstorms in Chapter 8.

The emphasis is also on relaxing. This is not easy in the charged atmosphere of turbulence, but you must settle down and make the most of life while keeping the pitch attitude within three or four degrees of what you want, and the bank within ten degrees of level.

Don't fight it. In normal en-route or approach turbulence, not of the thunderstorm variety, that would cover it too. There, of course, you have to maintain altitude, but in light-to-moderate turbulence an occasional slight adjustment in pitch attitude and power will handle altitude if it gets too far from what was assigned.

Probably the most "natural" thing a pilot does is sit on the edge of the seat and squeeze the wheel hard with sweaty hands when there's turbulence during instrument flight. The pilots who do the best job of flying under these conditions are the ones who sit farthest back in the seat, have the gentlest grip on the wheel, and sweat the least.

———————— *PARTIAL-PANEL BUMPS* ————————

If turbulence should be encountered after failure of the artificial horizon, flying is more difficult but still very possible. The fact that the airspeed will stay near the value for which the airplane is trimmed if the wings are kept level is a big help, and outlines the basis of the task: keeping the wings level. It takes a lot of interpretation to continually visualize the attitude of the airplane while flying in turbulence without the artificial horizon, but it can be done.

———————— *APPROACH* ————————

Turbulence behind, let's move on to some of the more sophisticated applications of our basic instrument-flying ability.

In the preceding chapter, in the discussion of partial-panel, we worked through an approach with the least in gauges. That is an emergency procedure, for time of need. Apply the basics to normal approaches, too. The interface between the basics and radio procedures is what converts instrument flying into a complete IFR flight.

One thing that must constantly be in mind in every situation is that the pilot's most important job is controlling the aircraft. If control is in jeopardy and the pilot tries to reach for the microphone, or look at the chart again to check a heading, then the trouble is multiplied. So while we have to plan around navigational problems during an instrument approach, any indication that control is problematic is a clear call to fly the airplane first and work on radio procedures when there is time.

Let's study an ILS approach first.

When we come to the outer marker outbound, or get close to the final approach course under radar guidance, we know that x number of things are going to have to be done to and with the airplane before we can walk into the airport office and tell the troops about that tight approach we shot.

First we have to slow the airplane down. When to do this? The important thing is to have a definite time and to accomplish the chore at that time.

If it is a straight-in ILS with radar vectors, they should tell us when we are getting close to the outer marker. The flight is usually transferred over to the tower frequency a few miles outside of the marker, and this is a logical time and place to reduce speed. Or, if your airplane is so equipped, the DME or the RNAV would suggest a time.

Each airplane is different, slowing down and descending is strictly an airplane-control item, and it is up to the pilot to develop a procedure that will result in the right thing happening in advance of the eleventh hour and on a methodical basis. In some cases, around busy airports with jets behind, you might need to keep that speed quite high on final. This can be done, and it's an item to practice. Certainly you wouldn't want to comply with a "130 knot on final" request for the first time while shooting a tight one to Washington National.

## HOW SLOW?

How much should we slow down for a normal approach? There are a lot of schools of thought on this. Some like to fly them fast; some like to slow-fly the approach unless requested to do otherwise. I like them fast. The instrument runways are seldom short, and most light airplanes are easy to slow down for a normal landing once you are contact at the published instrument minimums.

On a Twin Comanche I used to fly, I used 120 knots with the gear down as a normal speed for an ILS final approach. The controls were very effective at that speed, and application of full power for a missed approach didn't produce an airplane that was badly out of trim. In the twin, if an engine should fail (that always comes up), even during the first stages of a go-around, there would be no question about good directional control. And the higher speed got me down and out of the way quicker so the next fellow could have his turn. In my Cessna Skyhawk, I flew the ILS at 105 knots indicated, which is close to its normal cruise. Why not? It didn't have to be slowed down or trimmed, control response remained as it had been through the flight, and a go-around was exceptionally easy. I use 120 knots in the Cardinal RG, because that speed is okay with the gear down. And when flying any other airplane I try to look for a comfortable approach speed that is high enough to get the airplane down and out of the way quickly yet slow enough to allow extension of the landing gear at a normal time of my choosing. In a fixed-gear, I just let the airplane seek a comfortable speed on final.

——— *WHEN PUSH COMES TO SHOVE* ———

With the airplane sliding along on the approach, I've always been impressed with how every instrument on the panel must be watched closely, and how the navigational instruments add a little mental push and shove by taking on added importance and demanding a lot of the pilot's time.

For a basic example, let's talk about airplane control on a localizer (no-glideslope) approach. Pick up the approach after completing the procedure turn or being vectored to the final approach course. The airplane is flying level at the final altitude, on the inbound heading of the localizer, toward the outer marker.

There is where we really start learning about the actual winds aloft. And when starting down after passing the outer marker remember that the wind will likely change during the descent.

If the localizer needle indicates the airplane is drifting to one side or the other while maintaining altitude with the heading steady on the inbound, a correction is taken toward the needle. About 15 degrees is good for beginners if the needle has moved away from center at a moderate rate. If it crept away, use 10; if it sailed away use 20. After taking our example 15-degree cut toward the wayward needle, an active instrument scan between the localizer needle (to see if the correction is bringing it back toward the center, if it just arrested its departure from center, or if the needle is still moving away), the artificial horizon (wings level, pitch attitude correct for level flight), and the DG (steady on the correction heading selected) is necessary.

The scan needs to be active because if the correction head-

ing brings the localizer needle back toward center it is important to be aware of the rate at which the needle moves. If it
comes back in very slowly, then when it reaches center it
would be best to take off, say, only 5 degrees of a 15-degree
correction and fly with a 10-degree correction for drift. If it
comes back very quickly, the message would be to take off
most of the correction as the needle centers. This all accen-
tuates the basics of holding a heading. If you don't know and
hold the heading, you don't know what causes the needle to
move.

## OUTER MARKER, INBOUND

The balls which have to be juggled at the marker include
noting passage of the compass locater on the ADF and noting
the 75-mc marker receiver indication by ear or eye, calling
the tower to report the marker inbound, establishment of a
rate of descent by changing power and attitude (or lowering
the gear and changing attitude), and in some cases turning
the audio off on the marker receiver to restore a little peace
and quiet to the cabin.

What did I forget? The most important thing: keeping the
wings level so the heading won't change.

It's good to know an airplane well enough to be able to
change power to the letdown setting almost by ear—at least
to the point where you can set the power approximately and
then glance at the power instruments and airspeed momentarily to verify that all is in order. The microphone doesn't

need to be seen to be spoken to, and actually that call to the tower while passing the marker is a very low-priority item, to be accomplished after everything else is in good shape. If the marker-beacon receiver is so loud that you want it off, it should be something that you can turn off without looking at the switch.

Eye time is at a premium at this critical moment, and the artificial horizon needs a lot of it. Many an approach has been ruined because the pilot became preoccupied with all the other stuff and let the airplane bank and the heading stray, with the result that the localizer needle sailed away from center at a good clip. That begins the final part of the approach on an offbeat note.

## ———— FINAL APPROACH HEADING ————

Here I made a note to myself to pontificate on the importance of determining the heading required to track the final approach course and maintaining that heading. Then I backed away from that thought at least a little, because the heading *seldom remains constant* in a perfectly tracked approach during instrument weather, as the conditions associated with inclement weather often result in a shift in wind direction and/or velocity between 2,000 feet above the surface and the surface. When this happens, the heading cannot remain the same all the way in if the airplane is to maintain the desired track.

The emphasis is still on heading: Keep it steady on a

known value so that you can catch a wind shift quickly during the descent. Then any correction can be made on the basis of what the nav needle did in relation to a known heading, rather than on the ignorance that is bred by flying along with the heading varying 10 or 15 degrees either side of center.

Don't use the likelihood of a required change in heading during the final approach as an excuse for large heading excursions or for letting the navigation needle stray far from center. If the nav needle gets more than a couple or three dots away, it is time to abandon the approach and try again. It will not get far from center if the scan is active and if the mind keeps a running tab on the situation.

Fly an approach:

"There, 040 is tracking the localizer. Good. Now it is moving to the right slowly. Not much. Turn right to 055. That is moving the needle rather rapidly. Let it come back almost to center. Okay, all but centered, back left to 045 to track."

Note that the heading is always related to what it does to the nav needle. If you don't hold the heading there's no way to develop a picture of the relationship between the two things.

As you get closer to the airport, the maximum acceptable heading change should become smaller. If at any time it seems that a big turn is necessary, or the heading strays so far from where it is supposed to be that the nav needle starts to move from center rapidly, then it is time to call that approach a miss and start over.

## THE STRAIGHT AND NARROW

Heading management must tighten up toward the last, because the localizer course gets progressively narrower as the runway gets closer. It's like flying into a funnel. Localizer courses have a total width of from 3 to 6 degrees. The total course width of a typical localizer might be about a half mile at the outer marker. So if you are a quarter of a mile (1,320 feet) to one side of the center of the localizer course at the outer marker, the nav needle will show a full-scale deflection. At the approach end of the runway (the localizer transmitter is at the other end) the total width of the localizer is about 700 feet, so if you are only 350 feet from the center line you'll have a full-scale needle deflection.

The necessity for maintaining a heading is clear when an incorrect heading is applied to a localizer. If you are halfway from the outer marker to the runway threshold and at 105 knots groundspeed, a heading 10 degrees away from that which will keep the localizer needle centered will result in a full-scale deflection in only about twenty-five seconds. It is a precise business, and a little heading change goes a long way.

## A FINE POINT

On heading changes, a fine point is determined by what kind of airplane you fly. In some airplanes you can turn 5 degrees with the rudder and not disturb the bank attitude of the airplane much. That is, the wings can be kept level with little if any cross-control. In others, a little rudder at approach speed seems not to have much effect, and you have to bank for even the slightest correction. Try your airplane to see how it comes out.

I think that airplanes in which you can make a small heading adjustment with rudder are the easiest in which to track a localizer. When the localizer needle strays just a little from center you can move the DG 5 degrees with a foot, and have everything else remain relatively undisturbed.

Airplanes you have to bank to get a small correction aren't as easy to keep on the localizer, because the pitch-attitude status quo is more likely to be upset when you make a bank correction.

I have flown all the way down the localizer course without mentioning the airspeed and the altimeter. The airspeed will be okay if the power is right and the pitch attitude is correct. The altimeter, though, is a very important reference instrument, as it defines the minimum descent altitude if you don't have a glideslope and the decision height if you do. We'll be talking more about altitude later on, especially in Chapter 14.

─────────── *MISSED APPROACH* ───────────

Next comes the toughest moment of the approach, especially when you are doing it without a copilot, as most of us do. At minimums the pilot must look up, away from the instruments, and make a decision. If a runway is there and in the proper location, the message is to go on in for the landing. If a runway is not there, it's back to the instruments and square one.

Climb power has to be added for the missed approach, and then there's that big change in pitch attitude to get the airplane into a climb. The gear has to come up, if applicable, and the airplane probably has to be trimmed. All this while you're wondering "What next?"

There shouldn't be much acceleration error in the artificial horizon when you're changing to a missed approach, so you can just look at the horizon and bring the nose smoothly up to a normal climb attitude while adding power. Check the altimeter and vertical speed regularly to make certain a climb has been established, and also to verify the correctness of attitude and power (a glance at the airspeed indicator).

The things which probably make a missed approach the most difficult are the look away from the instruments to see if the runway is there, the disappointment if it is not, and the thought that the ground is so close and you are still on instruments and faced with a difficult task. It's a good thing to practice to perfection. And to keep practicing. Most of us don't shoot a lot of instrument approaches. We shoot even fewer when the reported weather is at or below minimums. So our chance for actual experience at missed approaches isn't too great, and we have to be satisfied that practice does indeed make perfect.

## *NEXT*

Shooting good ILS approaches should be considered in steps. First, the localizer needs to be mastered. It's not even bad practice to fly the localizer up high, level, without worrying about the descent. That way it's possible to learn the increase in sensitivity of the needle as you get closer without having the added distraction of the letdown. After the localizer is mastered, the descent at a given rate to the no-glideslope minimums can be added. Then can come the glideslope.

## *THE LETDOWN*

On a non-precision approach, one without the benefit of a glideslope, the letdown is quite a simple matter of basic airplane control and discipline. A good rate of descent at the proper airspeed to the minimum descent altitude is all that's required. Not one foot lower than MDA, either, unless the runway is in sight.

The glideslope, though, adds another localizer-type task to the approach. The descent must be made to correspond with the electronic path. The principles that worked on the localizer are applicable to the letdown on a glideslope, except that we will be talking of different instruments.

Just as a heading will track a localizer, a rate of descent will track a glideslope. The heading required is affected by wind direction and velocity, which are apt to change as the airplane descends. The rate of descent required is determined by the airplane's groundspeed (not airspeed), so the rate is also likely to change as the airplane descends and the wind aloft changes. Thus, while we can go at a glideslope with a predetermined configuration (gear down and approach flaps, for example) and power setting, there is certainly no guarantee that this will track the glideslope at the desired airspeed. And there is no guarantee that what tracks the glideslope as the approach begins will continue tracking the glideslope as the approach continues.

Too, just as we must know the relationship between a given heading and the reaction of the localizer needle, we must know the relationship between a given situation— power, attitude, airspeed, and rate of descent—and the glideslope. Swooping and dipping just does not work.

In light airplanes, speed on an approach is not usually critical (unless it is grossly slow), and we tend to adjust speed to help the controllers with the flow of traffic. ILS runways are much longer than necessary for light airplanes, so we are not likely to worry much about some extra knots on final. Making the corrections necessary to track the glideslope when the airplane starts straying is thus something that can be done with the elevators, with power, or with coordinated use of the two. It's a place where those classic arguments about controlling altitude with the elevator (or with power) can be hashed over for endless hours. Never mind, though. What you had best consider is that the pilot controls the airplane, and the pilot had best make a correction promptly when the glideslope needle begins to stray from center.

Say the needle begins to drop, indicating that the airplane is high on the glideslope. Lower the nose a little to recapture.

That will surely be effective. Or reduce the power a touch. That will also be effective. Personally, I think that the correction with power is a little easier, because it is measurable. If the airplane is a bit high on the glideslope when flying with 16 inches of manifold pressure, reduce to 14 to recapture the slope and then come back to 15 to track. Make a little pitch-attitude correction with the power change to keep the airspeed constant.


## NOTHING MAJOR

It is important to beware gross corrections, big control movements, and other dramatics on any approach. Smooth and basic flying is what gets the airplane to the runway. This is perhaps best illustrated on a clear day. Fly the airplane visually to the outer marker of an ILS. There, with the needles centered, set the airplane up in the rate of descent that should track the glideslope and fly the published heading of the localizer. Maintain both the descent rate and the heading until at the decision height (unless the airplane drifts far off and it would be hazardous to continue down to DH) and then note the airplane's position in relation to the end of the runway. Unless the wind is strong, the airplane will likely be close. Thus, relatively minor corrections, one that could have been made with your pinky, would have resulted in the airplane's flying out the small end of the funnel and onto the runway at the appointed moment. Remember this if you are ever tempted to make a wrestling match out of an instrument approach. Remember that doing nothing more than holding a heading and maintaining a rate of descent will *almost* do it.

True, in situations with strong wing shear some rather large changes will be required, but these are the exception rather than the rule, and the changes should be made smoothly when necessary.

## DON'T CRASH

A very important glideslope-related moment comes when the runway or the approach lights are sighted and the decision is made that the flight is visual. Most pilots will at this time fly below the electronic glideslope in establishing a visual approach to the runway. There seems to be a compelling urge to do something, to make some motion to mark the transition from instrument to visual flight. That something is usually a power reduction or a pitch-attitude change in the nose-down direction. This is unfortunate, because the electronic glideslope gives minimum clearances over everything and there is certainly no reason to go below it once the runway is in sight. When a pilot sees the approach lights and the runway, the best thing is to just continue flying the glideslope. The power, pitch attitude, and descent rate—everything—should remain at what was necessary to track the glideslope. It got you that far, and the glideslope will lead you on to a good landing point.

As an illustration of how pilots go below the glideslope, the National Transportation Safety Board noted that most IFR-approach accidents happen after the airplane is clear of cloud and on the visual part of the approach. It is only logical that the airplane would usually break out at some time before an accident; the point of the NTSB observation is that the visual part of an instrument approach is critical, and pilots tend to disregard guidance at this time. At least they do have to go below a glideslope to crash. If it's an ILS, the message is to stay on the glideslope after becoming visual; on a non-precision approach, the minimum descent altitude should be maintained until it is absolutely necessary to leave that altitude to land.

As we move from basic flying by reference to instruments to actual IFR operation, the primary thing to remember is that basics come first. There is a lot to contemplate about planning, weather, the system, and other facets of IFR flying, but we need to always remember the basics and spend ample time critiquing our basic ability and practicing to make it as perfect as possible. Too, when things go well on an approach, think in terms of its being a reward for a good job of heading and descent management. Those items and those items alone are what keep the needles crossed.

# 4

# ACTUAL INSTRUMENTS

An important step in learning instrument flying is using the rating after it has been earned. And sometimes it seems as if the rating is a lot easier to get than it is to use, primarily because the training system seems prone to avoiding actual situations.

For example, one day I noticed a fellow waiting for the weather to improve before starting from New Jersey to the Bahamas in a Cherokee Six. He had started from Long Island, and was en route with his wife to pick up another couple in Washington and then flee the winter. The forecasts had been good, but the weather wasn't living up to the forecasts. What was supposed to have been good VFR was actually at IFR minimums.

The pilot was in a quandary. He kept saying to his wife that they could file IFR and be out of the small bad-weather zone in just a few minutes' flying, but I could sense a lack of conviction in his voice.

The pilot never would file and go, and at the time it seemed questionable that he really had an instrument rating.

Finally one of the local pilots offered to take them home for the night. They accepted, and left the next morning in CAVU weather.

The pilot's overnight host later said that the man did indeed have an instrument rating. But it was brand new and had never been used. The proposed flight would actually have been as easy as they come, but the pilot just wasn't ready, willing, or truly able to tackle an actual IFR. It was of great credit to him that he realized this and decided not to make the flight. It was of discredit to the training system that he wasn't ready, willing, and able to tackle actual IFR when rated.

There has to be a first time, and someday this pilot will complete his instrument training by actually using the rating. Perhaps he would have that day if his wife had not been along. She didn't seem keen on the instrument flight, and some of her hesitation probably rubbed off. Or maybe he'll never use the rating. Too bad, and if that is the case I hope his instructor has a guilty conscience.

## ACKNOWLEDGED HURDLE

There is no question that the first actual instrument flight, or the first instrument flight in a while, is a hurdle. To many pilots the first actual IFR is as memorable as the first solo. I know that I remember mine as at least equal to it.

It was a short IFR flight, in a Piper Pacer. The purpose was to get through a bit of a front that was draped across the Allegheny Mountains. I had flown quite a distance in VFR conditions, and I had landed because VFR flying became impossible. The weather was checked, and it appeared cloudy but benign. I'm sure my hand trembled a bit as I filed that first IFR flight plan. In 1955 light-airplane IFR just wasn't the great outdoor sport that it is today, and I was doing what nobody else was doing that cloudy day. There were a lot of light-airplane pilots waiting at the airport for improvement in the weather that day. All eyes were on the fuzzy-faced kid in the Pacer.

I had gotten my instrument rating without ever penetrating a cloud, and my mind was awash with questions about IFR as I walked to the Pacer to commit my hide and its Grade A fabric to the mercy of the clouds. The unknowns outnumbered the knowns by a commanding margin. The only rationale I could muster was that conditions did have the appearance of being rather ideal for the venture. Besides, I knew that I would never learn how to fly IFR by watching from the sidelines.

The flight was an anticlimax. I was both nervous and excited to begin with, but when the airplane didn't melt as it entered cloud I felt some measure of relief. The air was smooth, and the temperature stayed comfortably above freezing. The Pacer ran like a top, and in a short time I was over the cloud-draped mountains and in VFR conditions on the other side. I felt the exhilaration that one might feel after parting with virginity. The mystery was gone.

The dramatics shouldn't have been part of that first flight. It should have been more routine. The fact that it was so special meant that if I had faced it at another time in life, later on, with family responsibilities and things like that, I might not have flown, just like the nervous Cherokee Six pilot.

## HOW?

How might we make actual instrument flying easier and more natural? The question is applicable both to new IFR pilots and to infrequent ones.

First the obvious things.

It isn't a good idea to use a brand-new or a rusty rating when there is strong pressure to deliver. The fellow in the Cherokee, for example, felt pressed by the fact that he had told people in Washington he would be along that afternoon. There was also perhaps some pressure from his wife. Even though she didn't appear overly eager to go, her husband probably knew that she was aware that the instrument rating and all those transistorized marvels in the panel represented quite an investment, and there would be some natural wonder about why it wouldn't serve as intended at the appointed time.

In my case, it was a matter of my getting to New York on time or joining the ramp-walkers waiting for VFR conditions. The latter would have been easier. There was some pressure to go, though. I had to get there. Had I picked a more relaxed atmosphere for my first actual IFR flight, perhaps I might have approached it with aplomb.

---

## RELAXED ATMOSPHERE

Rather than going for a command performance as a first actual IFR flight, or to break the ice after a period of inactivity, it is far better to fly with as little pressure as possible. Do it as if the day is a beautiful one for floating about the countryside and the flight is for the purpose of learning skills as well as for enjoyment and the basic challenge of flight.

If possible, pick a time when there isn't much traffic, and when conditions are thoroughly IFR but reasonably stable. Don't fly to go somewhere, just file a round-robin IFR and shoot a few actual approaches at the end of the flight if at an airport where multiple approaches are practical. I've often gone out on a Sunday morning, moved around IFR in cloud for a while, and then returned to shoot every approach in the book for my airport. With radar vectoring, it doesn't take long, and the practice is fine—much better than hood practice.

As noted earlier, it is a shame that actual instrument operation isn't conducted in all training, and any time a training institution avoids it, perhaps the pilot would be better off finding another place to fly.

## FAMILIARITY

A measure of familiarity with the environment inside cloud and/or rain is also important. The sensations in cloud can be special. The cloud form passing by can give a visual feeling of great speed. The water streaking the windshield and side windows gives a sensation not found when flying on a CAVU day. On the other side of the coin, there are some actual instrument situations where the airplane almost seems suspended, with little or no sense of motion. Whatever, the IFR pilot's world is inside. Don't let cloud or rain pull vision or thought away from the gauges.

Rain can be quite a diversion. Sound tends to draw the eye, and rain beating on the windshield is no exception. There is a strong tendency to look at the sound, especially when it changes. There is no time for that in actual IFR. A pilot must develop the discipline necessary to refuse to respond to any distraction.

Another example of a distracting time is when flying in and out of cloud. It takes a bit of concentration to stay with the instruments and not be drawn to an examination of the wisps of cloud whipping by outside or the cloud formation into which you are about to fly.

## HARD TO TAKE

I have heard reluctant instrument pilots say that it is difficult to accept the fact that it is normal to be in a situation where a handful of gyros and personal concentration on the instrument panel is all that is between them and deep trouble. Perhaps it is, but when flying VFR there's really little other than personal concentration and ability keeping the old devil at arm's length. And there are many other fields of endeavor where similar situations exist. A lot rides on one tire of a car when negotiating a curve at maximum allowable speed. When boating, only a half inch of fiberglass might separate sailor and shark.

## ———————— *THE OBJECT* ————————

When moving into actual instrument flying, confidence finally comes when we learn to approach IFR in the same manner as VFR flying. Some experience is necessary for this. The fact that there are clouds means little or nothing to the airplane unless those clouds are hostile, as in the case of icy or cumulonimbus formations. The pilot needs the same attitude toward cloud as the airplane has, and only actual instrument time will create that attitude.

Another beginning helper might be to fly the first few really actual IFR flights solo. Certainly passengers can be a diversion, and it goes without saying that a pilot is more relaxed when solo than with two or more eyes eagerly watching every move.

## PICKING WEATHER

Once a pilot feels comfortable with the airplane in cloud, it is still wise to bite off small chunks. Try to pick stable weather for the first few flights. Do be aware, though, that weather forecasting is an inexact science, and that you had better be prepared to fly every instrument flight in cloud, in turbulence, all the way. Accept the fact that the destination might be at minimums, as might the alternate, regardless of the forecast. That is the maximum task that one might be called on to perform, and it is always best to be prepared to do the most. If the weather is better, it is a pleasant surprise. And pleasant surprises are the only kind that should be allowed in instrument flying.

# PREFLIGHT ACTION

As we move into the real use of an instrument rating, acknowledge right off the bat that IFR flying is challenging. In fact, there is nothing more demanding or challenging in flying than the production of a smooth and precise actual instrument flight. A good foundation is absolutely necessary. The quality of preflight planning can actually make or break a flight.

A common conception of preflight work is that it is nothing more than dull drudgery. The image that pops to mind might be of a sea of pilot applicants locked in a testing room, poring over warped computers, trying to find the best answer in a multiple-choice exam that was designed to harass instead of teach, and filling out torturous little forms with headings and anticipated groundspeeds.

Try not to think of IFR preflight planning as drudgery,

though. Instead, try to consider it a stimulating exercise that makes the flight easier. A proper amount of planning will help avoid any number of booby traps that can foul up your basic instrument flying and make an adventure out of a trip that should have been routine.

The importance of planning can be easily emphasized if we follow that good practice of grading all flights. The poor grades most likely come when the unexpected creates a diversion. If the event had been properly planned, the unexpected would have at least been anticipated and would not have derailed the mind and caused so much trouble.

## THE MACHINE

Much of the preflight of our airplane is continuous. For example, I *know* that my airplane has all the required paperwork on board. The required VOR check is done and recorded on a regular basis, so the VORs are always ready for IFR. The transponder, altimeters, and static system are always kept current. The avionic equipment is kept in perfect order. No squawks are ever left over. The same goes for the condition of the instruments, and the airplane in general. If it isn't 100 percent at the completion of a flight, it gets fixed before the next flight. Instrument planning begins with knowledge that the airplane was put in top shape after last being flown and should be ready to go, with only a verification of conditions required before takeoff.

## THE PILOT

We will look at special IFR-related items for the airplane more thoroughly in Chapter 13, but it needs to be said here that just as the airplane should be a little extra for IFR, so should the pilot. I personally feel that there are times when a pilot is okay to fly in good VFR conditions but should not fly in IFR conditions. For example, if I have a bit of a head cold I don't mind flying on a pretty day when climbs and descents can be gradual and at whatever rate my ears will allow. But it is difficult to impose a 200-foot-per-minute rate of descent on the IFR system, and it shouldn't even be attempted. Nor should uncomfortable rates of descent that might cause ear problems and create a throbbing diversion from the primary chores. Any general feeling of malaise, whether from a self-inflicted wound at a party or just general blahs, is a sign to steer clear of real IFR. When you punch into bumpy clouds just before daybreak, and face the task of managing the instruments while the airplane bounces up and down, the message is plain that this activity is reserved for people who are awake and alert.

Don't read this as a suggestion that IFR is reserved for superpeople flying four-engine Boeings. All it says is to fly IFR in a good airplane and when feeling reasonably chipper.

## THE ENVIRONMENT

The next step is to check the weather. This involves organization, discipline, and a basic knowledge of meteorology.

A pilot who keeps up with general weather trends all the time is far ahead of the pack when preparing to fly. If you know that there was a cold front to the west last night, and if you look out the window this morning and note rather low clouds moving quickly from the south, you can begin putting things together before calling. The cold front is still to the west. Moist flow ahead of it. Strong flow? Look again. The trees are swaying—strong flow. The weather map in the paper or on TV will give some hint of the big picture expected for the day, and if you have a low-frequency receiver plus a continuous weather broadcast in your area, a thorough prebriefing is available as you shave or have coffee. Many times static from thunderstorms on the low-frequency has defined a day's problem for me before I made contact with the organized weather-information system.

While it is possible to go into a weather briefing without any preparation, the pilot who skips the preliminary effort is swinging with a short bat.

## THE PLAN

Before calling (or visiting) the FSS, determine what information is needed. I have a form with a little map on which to draw the synopsis (the location of the highs, lows, and fronts), with space for sequence reports and forecasts, and with boxes for winds aloft, Sigmets and Airmets, radar reports, and freezing-level predictions. I list the airports that report weather along my route, plus the destination and alternate, and then I let the form discipline my briefing. Personally, I always listen to the transcribed broadcast before calling, and often record information from it on the briefing form. Then I call the FSS for the rest, and I don't hang up until all the blanks have been filled in. My printed form has a place for the flight plan, too, and as much of it as possible is completed before calling. When I don't have one of the printed forms, I make a form before calling—on the back of an envelope, anywhere.

## ───── HELLO, MY NAME IS . . . ─────

It is important to get the telephone briefing off to a good start. Establish that you are a pilot, the person responsible for a flight, and outline the desired information: "Good morning, I am going IFR from Little Rock to Wichita, below 10,000, will need the synopsis, the Little Rock, Fort Smith, Tulsa, and Wichita weather plus the Wichita forecast." That tells the person that you know what you need, and it hopefully avoids the kindergarten delivery of weather information that is often inflicted on pilots who sound disorganized and helpless when calling for information.

After the briefer gives you this first load of facts, follow up by asking for the alternate weather and forecast if required, plus the winds aloft, radar summary, Sigmets and Airmets, and the freezing level. Never fail to ask for all of those items. Maybe it is summertime and the freezing level is above the ceiling of the airplane, or maybe it is below zero and the closest thunderstorm is in another hemisphere, but the discipline of asking every time is important. You might even get useful information. In the summertime, cumulus clouds start becoming cumulonimbus as they build through the freezing level, so information on that *is* something worth having. And I've seen thunderstorms when it was snowing, so never count them out.

If the list is complete and if the briefer answers the questions, the next item is to file the flight plan. Three flight-plan items probably can't be completed before calling—the altitude request, time en route, and alternate—but these can be added at the last minute. Somehow using one call for a briefing and calling back to file a flight plan seems wasteful. Also, with the lines as busy as they are, it might be noon before you got through twice.

## WAIT A MINUTE

Did I miss something? Well, that much-heralded and holy savior of pilots, the go/no-go decision, was not discussed, and it might be argued that it should have been. I find, though, that the best way to plan IFR operations is to think only in terms of "go."

The reason it is best to plan only for go is that you can't make an intelligent decision while talking with someone on the telephone. The object of the call is to gather the information and file the flight plan. Any decision is best made in a quiet and reflective moment, after all the available information has been recorded and assimilated. If a pilot starts trying to make a decision on the phone, often as not the pilot will wind up mumbling in parables and virtually asking the briefer to help decide. A briefer can always find at least one reason for you not to fly on a given day, and some of them make an art form out of terrifying pilots who seem reluctant. Consider the briefer only an information source. The pilot is responsible, and only the pilot can make intelligent decisions about the conduct of a flight. And if a pilot hasn't had the wisdom to develop more than a basic knowledge of meteorolgy, heaven help him or her in IFR *or* VFR flying.

## WHY NOT?

The two most frequently found reasons to scrub an IFR flight are thunderstorms and ice. These will be discussed fully in Chapters 8 and 9 respectively and we'll look at the go/no-go decision in relation to them at that time. For the purpose of this chapter and the next, we'll assume that neither is an operational consideration along the route of flight in question.

When a flight is of any length, the go/no-go decision has to be tempered with respect to the fickle nature of weather and the usual inaccuracy of forecasts. Experience will come to suggest, perhaps later rather than sooner, that reliability comes from making the initial go/no-go based on the conditions to be expected over the territory covered in the first half-hour of flight. The decision that is made before flying is that it is okay to start out. After takeoff, the go/no-go decision is continuously examined, and the flight is extended, terminated or diverted to an alternate based on actual conditions ahead. A pilot should never feel bound by a go decision. An instrument pilot simply must be a big enough boy or girl to play the game by flying it one mile at a time while keeping a wary eye on the miles ahead.

## GASSED

Another item to complete before flying has to do with fuel. If you know enough about the airplane to be flying it, you should be able to make an accurate determination of the fuel that will be required for a flight of any given duration. Once the calculation has been made for the flight to the destination and the alternate, make sure there will be an hour's fuel on board at the completion. That is 15 minutes over the legal requirement, but every pilot owes himself an extra 15 minutes. The tanks might not be quite full to begin, there might be a delay getting off, some vectoring, or some other factor might increase consumption slightly.

Once the fuel calculations have been made, make a mental note of groundspeed required to complete the flight on this amount of fuel. This is a magic number. Next, compute an estimated time en route to the first fix along the way, based on this groundspeed. Then, a late arrival at that first fix or any indication that the groundspeed is below the magic number is a demand that the fuel reserve be recalculated. Any deterioration of groundspeed later on means the same thing. If, on recalculation, it appears that landing at the alternate will be with less than an hour's fuel on board, the flight must have a new plan, a new destination and/or alternate. When my hour's reserve is going to be close, I always write the necessary groundspeed on the flight-plan form in rather big numbers, as a reminder of its importance.

Remember, too, when using published information in planning flights, that the data given on range in miles are absolutely useless unless the wind happens to be calm. That is highly unlikely, and what counts in every flight is time. The fuel runs through the pipe by the clock, not by the mile. Don't be suckered into thinking that a 500-nautical-mile flight will be comfortable because the book shows a 750-nautical-mile range at the chosen power setting and altitude. Planning must be based on endurance, with a generous allowance for taxi, takeoff, and climb, as well as for variations in fuel flow and any other contingencies that come to mind.

While planning, don't fail to determine the power setting for the flight. This is something that can be done in advance, so it should be done in advance. When you level off, you have the settings, and you know how much fuel goes through the pipe at those settings. In using the pilot's operating handbook and the power computer of my Cardinal RG, I've been able to predict fuel use quite precisely—often to within a half gallon for a flight.

We'll again consider fuel in the next chapter, on in-flight planning, but do remember not to slight preflight fuel planning. And there is no kidding allowed, because kidding leads to dry tanks and to silence.

## FLIGHT LOG

What about a detailed flight log, prepared in advance? I usually fill out a very simple one when looking up the airways for a flight. The list is of each station and the distance between stations. Once en route I record the time over each and compute the groundspeed. With DME I might not compute the groundspeed for every leg, but I would for some just to check, and I would always list the time over every station, to have something to build on should the DME fail.

There is certainly more than one way to keep books on a flight. Some pilots like elaborate logs, including distance flown, distance to go, the bearing of each airway segment, and other items. Some pilots keep up with their en route business on a blank piece of paper. Either way is fine. Just make sure you have a system. I think pilots enjoy inventing their own forms, and the jiffy printing services of the world make it easy for each of us to be in the IFR-form-printing business at a nominal cost.

## SID?

Sid who? Standard Instrument Departure, that's who, and while still planning the flight don't fail to look at any departure procedure for the airport in question. If there are published SIDs, the flight plan should have been filed along one of them. If not, chances are the clearance won't come back as filed and there will be a moment of silent confusion when the clearance is received. Equally important, the approach plate for the departure airport should be scoured for any instructions on how to leave the area, and this should be studied and understood in advance.

The departure procedures on an approach plate are usually for terrain or obstruction avoidance, and not heeding the printed word could lead to a rock-strewn hillside. Jeppesen includes these procedures, where applicable, on the plate with the airport diagram, at the bottom, in rather small type. I keep a complete set of WAC charts, and when a question about high terrain or obstructions is raised by something on the approach plate or en route chart—a special departure procedure or high minimum en route altitude, for example—I study the WAC chart for an understanding of the reasons. And at times I've noted that a different airway, involving only a few extra miles, will take a flight over lower and more friendly terrain than the primary and direct airway between two points. That is always worth a few miles.

## ROUTE SURVEY

A little route survey is a good windup, and is quite useful in setting the mind for the task ahead. Trace the flight on the chart from start to finish with your finger. Mentally position your weather information on the chart as you go, and visualize what it should be like along the way. If MEAs are high, check why. Note the availability of alternate landing sites. If the weather is really grungy, the airports with a full ILS take on special importance. They might well be the only true havens. A route survey is especially important when flying in an area that is not familiar.

## THE AIRPLANE

Some items of the preflight inspection become extra important when you're checking the airplane before an IFR flight. In addition to the usual good things, take extra care in looking at the static port or ports, because these affect the pressure instruments. Peer into the pitot tube and check the radio antennas, too. As things are stowed, make certain the baggage door is closed securely. If it comes open in flight, you'll almost surely have to land to close it. That might mean an extra instrument approach. Other things that might prompt a return in VFR conditions—coattails hanging out the door, opened oil-access doors, loose and banging gas caps, for example—can become major projects on an IFR flight. It is quite accurate to say that you are placing more faith in the machine for an IFR flight than for a VFR flight, and that faith calls for an extra measure of attention during the preflight inspection.

## ON BOARD

There's no question that a light airplane isn't an ideal place to do the type of work that is necessary during an IFR flight. Ideally we would have a large table on which to spread the charts; actually that large table must be compressed to the size of a kneepad or lap-held clipboard in the airplane. A little organization is necessary to make the most of this confined work space.

We all know that it is easier to fly any flight without passengers than it is with passengers. This phenomenon is even more pronounced when flying IFR. The empty seat to the right that can be so handy for books and charts is full, and there never seems a place to put anything. We thus need to devise a system for getting pilot, passenger, and charts neatly in place, with charts readily available. If possible, the system should not involve the passenger at all. Unless an active pilot, a passenger is best left completely in that role.

Most high-wing airplanes have an advantage over low-wing airplanes when you're arranging stuff for an IFR flight. In a high-wing with a door on each side, I get the passenger(s) in place and belted, and then stand outside the airplane and arrange all my little charts and papers before boarding. The Jeppesen books are stowed or stacked on the floor between and under seats. The proper en route charts are folded and placed in the clipboard, which is put atop the panel while I get into the airplane. In a low-wing, my system is to stack things on top of the panel (if it is large enough) while I board and then get everything organized after everyone is in and strapped. The main thing is to have a plan, to do as much as possible in advance, and to completely assume the role of aviator (as opposed to stewardess) once any passengers are buckled up and settled.

## ——————— *HUSH IN THE GALLERY* ———————

One other word about passengers. They must understand that IFR flying requires a certain degree of concentration, and that they should speak only when spoken to. Give some advance word on the methodical nature of IFR flying, so they won't automatically assume that the pilot isn't quite with it because of continuous references to checklists and charts. After that, retreat into a shell and care not what the passengers think. Every IFR airplane should have a good headset on board, preferably with a boom microphone. In addition to its other, obvious benefits, this device is useful in shielding pilot from passengers.

## ——————— *THE OBJECT* ———————

The object of pre-start and pre-takeoff work in the cockpit is to get as much done as possible before flying away. Make certain that everything that can be put in place is in place. The whole purpose of a checklist is to command the discipline necessary for the pilot to *know* that everything is just right when the throttle is advanced for takeoff. Checklists aren't for sissies, either. They should be used religiously in IFR operations.

Some checklists don't specifically cover all the IFR items, or they don't cover things in a logical sequence. In this event, make your own. The list should include every item, including things like setting the transponder and turning it on, putting the autopilot heading bug on the first heading to fly, setting the navigation radios, setting the departure-control frequency in the number two radio, and turning the pitot heat on. Get everything that you might need turned on in advance. The checklist should be arranged so that as many items as possible are handled before taxiing with only the last-minute things left for the runup position.

## THE CLEARANCE

Many different forms of clearance shorthand have seen the light of day. Clearance shorthand is like the flight log—to each his own. If a pilot gets all hung up over memorizing and using some specific form of clearance shorthand, he or she might well draw a blank over what symbol to use for one thing that is said and thus miss the rest of the clearance. Just write it naturally, using whatever abbreviations come to mind. Also, when copying a clearance do not try to make sense of the instructions as they come to you. If it is an involved clearance, that means it won't be as you filed the flight plan. That, in turn, means you will mentally question it. Resist the temptation to think of questions during the reading. Write it down, putting all the mental effort into catching the spoken words and making a record of them. Then read it back. Then read it again and make sense out of what it says.

Make sure you understand a clearance before flying, and make any necessary changes in navigational radio settings before takeoff. If, for example, your filed route was over Yardley and then outbound on the 276 radial, but your actual clearance is to intercept the 152 of Yardley and fly that to the Gdork intersection, and then to do other mysterious things, get it all straight before takeoff. Set the radios in accordance with the clearance. If this isn't done, that difficult period in basic instrument flying immediately after takeoff is likely to be confusing as well as difficult as you try to fly and figure out a routing at the same time.

I read not long ago of an IFR accident in which a pilot flew in exactly the wrong direction, 180 degrees out of phase, and finally lost control of the airplane in the confusion that ensued. Perhaps he was turned around to begin with, and once this was called to his attention he fixated on the navigation problem and forgot all about flying the airplane.

## READY FOR TAKEOFF

If the preflight effort was effective, everything will truly be ready for takeoff as the airplane moves out onto the runway. The airplane has been checked and is in good shape, the pilot is okay, every setting that can be made in advance has been made, the charts are at hand and are understood (and the approach plate for the point of departure is readily available in case the need to return and land should arise), the clearance is on paper and is understood, and the weather is suitable for the launching.

Something to watch: Many of us do well at all this planning for the first flight of the day, but regress as the day wears on. It shouldn't be this way. The second leg (and any subsequent leg) of a day's flying should be planned just as carefully as the first.

Take off, too, with the thought that planning does not end when the airplane leaves the ground. That is actually when the planning challenge really begins. Up to this point, time has been expandable as necessary to accommodate planning. Once you are airborne, there is a fixed amount of time available to do all the necessary things.

# 6

# IN-FLIGHT

This first item isn't quite off the ground, but it is close enough, and it is first. After the engine spins up at the beginning of the takeoff roll, I always check a few things. If there's time, I look at the rpm, manifold pressure, fuel flow (if applicable), vacuum reading, and alternator output. And I usually find time. My theory is that if the engine and accessories accept the acceleration to full power, they are likely to keep on working for a while. And if something in there is weak, looking for a time to malfunction, perhaps the increase to full power will expose that weak link. Needless to say, an indication of anything short of perfection would be a mandate to abort the takeoff and seek mechanical aid and comfort.

## PUNCHING IN

In speaking of basics earlier, the importance of concentrating on flying during the first few moments was stressed. Have a plan for the way it will be done and things will go extra smoothly. Unless there is a compelling reason not to do so, I like to climb on runway heading at full power to at least 500 feet above the ground (1,000 feet is even better) and then go to cruise climb before making any turn. If flaps are used for takeoff, they are left extended to this transition altitude. The gear is retracted normally, but I avoid doing anything else. It is my time with the airplane, a time of introduction.

There could be one variation on normal procedures, too. If there is a pitch change or a marked sinking spell with flap retraction on a particular airplane, it is good to take a hard look at the necessity for flaps on takeoff. Unless they are called for on an absolute basis, it might well be best to start out flapless and avoid any pitch change and sinking spell in those important first minutes on instruments.

Once in cruise climb with the initial communications out of the way, it is time to start gathering information to use in the planning process. In the weather briefing you should have gotten temperature-aloft forecasts along with the wind forecasts. Check their accuracy on climb. A really bad temperature forecast would suggest that all other forecasts might also be off by a good measure. Note the rate at which the temperature drops as you climb, too. If the temperature drops rapidly with altitude, that is a sign of unstable air. In the summertime this could suggest thunderstorm activity later in the day if yours is a morning flight. If the temperature increases with altitude, that's a sign of stability. There is not likely to be much bumpy action, but the surface wind and the wind aloft are probably far apart, and the ceiling and visibility might well have been low. Keep a continuous tab on those temperatures to gauge the accuracy of the forecast, and also to know where warm air is located in the event that ice should become a problem during a flight.

## WINDS

You can get some feel for winds aloft in a climb even without DME. If the airplane climbs better than usual at some time during the ascent, perhaps it is climbing into an increasing headwind at that point. This situation does result in a momentary boost in airspeed and climb until the airplane adjusts to the new flow. Remember the altitude at which it happens; above it you might go very slowly, but if you could fly below that altitude the headwind might not be so bad. On the other hand, pronounced sagging in climb might indicate an increasing tailwind. If there are mountains around, flow over these could be the cause for better- or worse-than-usual climb, as could cumulus development, so any variation in climb can't be taken as absolute word on anything—but it is surprising how often it portends ground-speeds to come.

The clouds tell some tales as we climb. If there was any indication of layers or tops during the briefing, be curious about the accuracy of the prediction. Continuous turbulence in clouds means they are cumulus types, unless there is some wind-related turbulence from terrain. Was the forecast such that cumulus clouds might be expected?

All this curiosity about the elements shouldn't be allowed to detract from the flying, but try not to neglect it. Once the airplane is off the ground it becomes a much better weather sensor than anything possessed by the National Weather Service, and the pilot should be keenly interested in the airplane's surroundings.

All the information gathered before takeoff was for the purpose of deciding that it was okay to start the flight. By the time the flight is over a wise instrument pilot will have collected a few more complete sets of data. What *is* actually happening along the route and at the destination and alternate is a lot more important than what *was* happening or what was forecast to happen. Whenever there is new information, get it. It isn't necessary to neglect basic instrument flying to do this. Just budget time properly.

## KEEPING BUSY

When moving along en route on a routine IFR flight, things don't jump at you and demand attention as they do in other phases of IFR flying. Someone prone to be a touch on the lazy side might even while away the hours immersed in boredom and idle thoughts. That is survivable some of the time on some flights, but the time often comes when such apparently mundane busywork as checking groundspeed and studying charts can help avoid thrills and adventures.

## GROUNDSPEED

In the preflight deliberations an estimated groundspeed was calculated. If the speed goes below the estimate, the fuel reserve will have to be recalculated. This becomes an important chore on the first leg. Add the estimated time en route to the time off for an estimated time of arrival at the first fix, and then keep a running tab on how things are going as you progress toward that fix. If there's DME on board, fine, use it. If not, a second VOR can be used for intersections along the way. Certainly intersections are not precise, and you wouldn't want to change an ETA based on the time over one intersection, but you can surely become suspicious of a gross error in the winds-aloft forecast if times over intersections are far off what was estimated, based on the winds-aloft forecast.

The ADF can be very useful in these initial groundspeed deliberations as well as at other times. Nondirectional beacons have proliferated, and if there's one on or along an airway, use it as a fix. Some broadcast stations are shown on Jeppesen charts, and these too can be used as fixes in calculating groundspeed. These positions would be at least as precise as anything you'd get with VOR cross bearings so long as the station is close to the airway.

The greater the distance to the first VOR along the airway, the more important it is to work at an initial groundspeed verification before reaching that fix. When I had a Skyhawk, first legs from my home base were almost all over 100 nautical miles. If the wind was 40 knots instead of 20, I needed to know all about that before the first VOR was reached, because an extra 20 knots on the nose was the difference between possible and impossible on a lot of trips. It is less critical with a faster airplane or with a lot of fuel, but the sooner you know the groundspeed the better in any airplane.

## WIND, WEATHER

Another reason for nailing down groundspeed early is to get a reading on the accuracy of the synopsis. If the wind forecast is badly in error, that means they didn't have the pressure systems correctly evaluated and the real weather map is probably somewhat different than the one you have in mind. A badly missed winds-aloft forecast should be a clear signal to dig carefully into all available weather information to see if the switch is going to have any pronounced effect on flight conditions. It'll likely mean a recalculation of fuel reserves, too, because, alas, the unforecast wind always seems to be from ahead.

## TRAFFIC

Most of our IFR hours are actually flown clear of cloud. This brings with it a total responsibility to look for, see, and avoid other traffic. It is a serious mistake to let the IFR busywork ruin the scan for other airplanes, and it certainly isn't necessary. The way to look for airplanes is with a methodical search pattern, and this can easily be maintained while flying the radials and altitudes and doing the necessary bookkeeping. This is a good time to use an autopilot, too. Let it track the radial and hold altitude (if equipped with that feature) while you divide time between looking for traffic and taking care of other IFR chores.

Controllers will call traffic for IFR flights, but this is done only on a workload-permitting basis, and it is nothing to depend on. Sometimes they call the traffic and sometimes they don't.

## TALKING AND LISTENING

Much IFR communicating is done en route, so this is a good time to examine the techniques of talking and listening.

The words said as we fly along are hardly conversational; they are more a series of grunts and acknowledgments. This becomes quite obvious when someone, pilot or controller, says something out of the ordinary on the frequency. It is difficult to understand words other than the magic ones that are used routinely. An example came one day, a pro-football playoff day, when I was flying over Texas. The Dallas Cowboys were playing, and pilots seemed to think controllers would be listening on their transistor radios, while controllers seemed to think that pilots would have the game on the ADF. In the course of flying for several hours in the area, I must have heard the question "What is the score?" asked a half dozen times. Not once was it understood on the first asking. Not even in Texas, with the Cowboys in the playoff.

What this tells us is to speak in a predictable fashion. Use the magic words. This is easy to learn just by listening to how others do it, especially airline crews. They fly more than most of us, and they have learned better than anyone the art of communicating with the controller. Their transmissions are very brief—no excess verbiage—and they speak more effectively than general-aviation pilots.

A small example: When cleared to descend from 6,000 to 4,000 feet, the general aviation pilot is likely to say: "Roger, Cessna 030 descend to 4,000, I'm, ah, out of six."

The airline pilot might say: "Out of six for four, American 330."

The airline version uses fewer words.

Why do airline pilots often put identification last instead of first, as we do? It could be that this is a more effective way of communicating, because the pilot's repeat of the clearance follows the controller's reading of the clearance without the aircraft's number coming between. Either way, identification last or first, the important thing is to say the message clearly and in as few words as possible.

## THINK

If what you are about to say is not routine in nature, compose a little speech in advance. You call a controller because you want something, and the more clearly and concisely the request is stated, the better. For example, they generally expect pilots to stay at one altitude for the en route portion of a trip. They do not *expect* you to call and request a different altitude halfway between here and there. So there is no anticipation of what you are about to say, and if you say it in a confusing way you might well expect to draw a blank. I heard the following one day:

"Ah, Center, ah, this is 40 Romeo Charlie, I'd, ah, like either a higher or a lower altitude."

"Say again, 40RC."

What is a controller supposed to do with something like that? The person on the other end of the line is supposed to be pilot-in-command of the aircraft, he seems to want to fly at a different level, and yet he doesn't know whether he wants to go up or down. Some command. If the pilot had thought, a more concise request could have been phrased: "Center, 40RC requesting higher altitude due to turbulence."

Less-than-concise communications can come from the ground, too, with controllers using different phraseology in different parts of the country. For example, a controller recently transmitted the following to me: "Zero three zero is cleared to Woodstown only, maintain 7,000." I thought he said "cleared to Woodstown holding," because I had never before been cleared to somewhere *only*. After listening to the controller clear a couple of other airplanes in a similar manner, I caught on that what he was saying meant that Woodstown was the clearance limit.

Basically the air traffic controller either clears us to do as we wish (as filed), gives an alternative, or tells us to do something. Relatively few words are used for a high percentage of communications. *Cleared, turn, right, left, climb, descend, cross, hold, approach,* the terminology for airways and facilities, plus numbers and place names account for a large portion of the words that they say to us. There is certainly nothing complicated about that limited vocabulary, but at times we still tend to misunderstand or misinterpret the word from the ground.

Managing thought processes helps in understanding the other person. Consider anything the controller says in the same manner that you consider a clearance. Listen to the entire message, then decide whether or not it was understandable and made sense.

If, for example, the controller says: "November 34030 is cleared to the Aramb intersection, to hold northwest, left turns, one minute legs, maintain 6,000," don't stumble over "Aramb" and miss the rest of the message. Get all you can and then ask for a repeat of the intersection name. That is a lot better than becoming derailed right off the bat and having to do the whole thing over again.

Thinking of a communication as a total message, with the finished product contemplated for a moment before acceptance or rejection, is a good practice. By that I don't mean to give the controller a fast "Roger" and keep on flying whether the message was understood or not. Rather, consider the transmission for a moment, in context, because if there is confusion in one part there might be clarification in another. This can be very helpful when a controller happens to use a bit of nonstandard terminology. "Cleared to Woodstown only" was understandable after a moment of thought. If one said "Cleared into position and hold for release," the word "cleared" might charge the mind with taking off, as in "Cleared for takeoff," but the rest of the message certainly clarifies and shows that what he really means is for you to taxi into position and hold.

Too, if you have trouble getting the message, consider the possibility that it might be caused by your questioning things in the transmission before the controller is finished. There is just no way to hear and interpret a message when the mind is charged with a question.

It is a cinch, too, that audio quality in some airplanes leaves a lot to be desired. This can be hard to improve, and if a new cabin speaker doesn't make the words clearer, headphones would be the only answer.

As you fly around the country, there will be some inevitable clashes with accents or even colloquialisms in language and different regional procedures. The key is to always get it straight. Never fly on with doubt about what the controller said.

Communication should be an easy and natural thing. It takes very few words to get through an entire IFR flight. Brevity is of the essence. The airline crews are aviation's best communicators. Listen, and imitate them.

## CHECKING WEATHER

As the flight progresses, get each hour's weather reports along the way. It is pure foolishness not to do this. It takes but a few minutes, and current weather information is all you have to use to avoid surprises. The decision to continue is based only on aeronautical factors, too. The IFR flight must be conducted in an isolated context, free of wishful thinking. If the fuel reserve is in danger of compromise, or if the destination isn't up to minimums when you are a couple of hundred miles away and the alternate is looking shaky, the importance of "being there" deserves not a thought. What counts is conducting the flight within the framework of conservative IFR principles. The greater part of the framework is defined by weather conditions, and there's no way to stay in bounds if you don't continually check the weather and heed the messages.

Remember, too, that a very important part of your mental weather picture is knowledge of where the weather is good. If a problem should arise, where could you take the problem for solution in the best possible conditions?

## WHAT IF?

Time now for some en route puzzles. Two hundred miles to go and the destination weather is 100 overcast and a quarter of a mile visibility two hours after it was forecast to lift to 500 and 5. The alternate is just barely hanging in there with 200 and ½ even though it is forecast to be 800 and 5. What to do? It is tempting to respond to such a situation by asking for a "new" forecast, without giving much thought to the fact that the new forecast will come from the same place that produced the old, bad forecast.

When things unfold in that manner, it is clearly *not* the time to bet on forecasts. It's time to study actual conditions and put together a picture of *what is really happening*. In such a case, I'd start looking for a place between myself and the destination that had decent-enough weather for an approach and pit stop—especially if the original fuel calculation was at all tight.

Continuing toward weather that is worse than forecast, regardless of what it is supposed to do next, is a bad deal unless the airplane is literally awash with fuel. Have enough to go to the destination, accept the busted forecast, go to the alternate, accept the busted forecast there, and then go somewhere that has weather that you know is good enough for an approach.

## ALTERNATE

We give a lot of thought to selecting alternates, and I suppose we do this because the alternate is part of the regulations. We *must* consider one for the flight plan, and it must be done according to the rules and forecasts. Given the inaccuracies of weather forecasting, though, alternates are best filed with a grain of salt. I certainly never consider that the filed alternate is cast in stone, and on a touchy day the true alternate in my mind might change half a dozen times during the course of a flight. One of my favorite alternates is the last airport with an ILS approach before reaching the destination. Such a place is more likely to have minimums than an airport served by a VOR or NDB approach, and before passing that last ILS before the destination I like to recalculate reserves, recheck weather, and make sure a continuation to the destination really is a viable proposition.

The IFR situation is often as not one of flying from relatively good weather into inclement weather. Keeping a close tab on that bad weather as we move toward it is important, and if things are really on edge, the best of all deals might be a fuel stop before flying into the area of bad weather. With plenty of fuel, the pilot has time to think and plan. And act. With minimum fuel, decisions are as likely to be influenced by the fuel gauges as by aeronautical and meteorological considerations.

## ATIS

As the distance to the destination grows short, it is customary to listen to the automatic terminal information service, if the airport has such. This gives word on the approach and runway in use, in addition to meteorological data. It also tells all the horror stories about 150-foot cranes operating in the vicinity of the airport. When is the best time to listen? I always tune it when in range, before things start to get busy. Some pilots wait until the center controller hands the flight off to the approach controller, and then listen before calling on the new frequency. This does insure getting the current information, but it doesn't give a lot of time for contemplating the approach. Listen early and there's time to plan the arrival. If the approach in use changes before you get there, well, you won't be any worse off than if you had checked it later rather than sooner.

The current weather is important only as a point of reference. If conditions are above minimums, you are probably going to shoot the approach anyway, and the operation should be conducted with the same concentration regardless of weather. It sure doesn't hurt to have the current weather, but don't bet on its being correct. I've anticipated good weather from an ATIS only to wind up shooting a really tight approach more than once.

One weather item that is important is surface wind. Make special note if it is markedly different than the winds aloft. This means there will be some wind shear on approach.

More important to planning than weather is the word on which approach and runway is in use. Good preflight preparation included a look at the big picture, but the real final details can't be planned until the actual approach in use is known. Then study the appropriate approach plate. If it is a non-ATIS airport with more than one approach, ask the controller which approach will be used before flying into the last phase of the flight.

## WHAT'S IMPORTANT

The approach plate is littered with information, and you can't memorize it all. And, really, only one item must be absolutely committed to memory: the minimum altitude to which you will descend on the approach. This is the minimum descent altitude (MDA) for a non-precision approach (no glideslope) or the decision height (DH) for a precision approach. This is a critical value, and disregard (or misinterpretation) of it is closely related to a majority of the serious accidents that occur in instrument flying. Descending below the MDA or DH without the runway in sight has proven to be the most hazardous thing the general aviation pilot does in IFR flying. The MDA or DH should be etched in mind before starting an approach, and it should be taken as an absolute discipline.

The bearing of the final approach course is also a primary approach item, but we have crutches here and don't really have to rely on memory except perhaps during an ADF approach. On a VOR approach, the number is set on the omni bearing selector and there is plenty of time to compare the setting with the chart and double-check it. On an ILS, the bearing is published, and for reference it can be set on the omni bearing selector. (When using a horizontal situation indicator, it must be set.) On an ADF approach, the bearing can be set if the directional gyro or horizontal situation indicator has a heading bug. Otherwise it must be remembered.

The missed approach is also an item to note and remember. Keep the basics in mind—climb straight ahead to 2,000, for example. If the missed approach involves a turn, memorizing the direction of turn becomes very important, because it is often *away* from some obstructions. Turn the wrong way and you'd be going toward those obstructions. It has been done, with disastrous results at times.

I also like to position obstacles or high terrain in my mind. The route survey before takeoff should have provided an overview of terrain, so we fly with basic knowledge of where the high stuff is. In looking at the approach plate to be used, the question is in relation to where the highest obstacles are around the specific approach and around the airport.

## TWO WAYS

Here we'll consider the planning of non-radar and radar approaches separately.

When making a non-radar approach—meaning the controller is issuing clearances without vectoring the flight—the only protection is in flying published routes and altitudes. When you learn of the approach in use at an airport that is bereft of radar services, note the available published routes carefully and plan the arrival accordingly. If the controller clears you for an approach when you are forty miles away, don't get in a hurry to descend. First, determine that any descent will be on a route on the chart and then descend only to the altitude shown on the chart for that route. Fly it methodically and don't cut any corners.

When you are in radar contact, the trip to the final approach course will likely be vectored, with altitudes assigned by the controller. The altitude given to you will be the minimum vectoring altitude or higher; MVA provides the legal minimum above the terrain or obstructions.

At first glance this vectoring business seems by far the easiest way to maneuver in preparation for the approach: Someone else tells you which way to point the airplane and the altitude to fly. We shouldn't turn into a vegetable during such an approach, though. Keep up with position, and make note of high terrain or obstructions and the airplane's position in relation to them. The pilot is responsible for the flight, and the pilot who completely relaxes about position and altitude when being vectored is hardly reflecting much command ability. I like to at least have World Aeronautical Charts on board to use in noting position in relation to other things when being vectored. If the country is flat and there are no

TV towers, no sweat. A glance at the chart shows that all is well. If it is mountainous, or if there are big towers, it's nice to confirm that the vectoring is being done away from or above these items. In case of radio failure, too, you would know the location of the briar patch.

When approaching a large airport, make note of any airline jet aircraft ahead if you are between layers or on top. Wake turbulence settles, and we do operate at a disadvantage, because procedures usually bring jets in above light aircraft; controllers clear the jets to descend after they have passed us. Thus, the big airplane will lay its wake right through our altitude. If you see a big one and have an idea for wake avoidance, tell the controller. A heading 10 degrees either side of that assigned might be just the ticket. If there is any question in your mind about being in the vicinity of the wake of one that passed through recently, at least slow your airplane to maneuvering speed. I've had two wake encounters in busy terminal areas. Both were at a right angle, and there was quite a bump.

Jets almost always make ILS approaches, so we get some help on wake-turbulence avoidance from the glideslope. The fact that the wake settles means you should be able to follow a jet with standard separation without encountering its wake. I always add just a little more in my favor by staying one dot high on the glideslope when following a large airplane.

## A BREATH OF O₂

Here, you might do something for the old brain and bod before tackling the approach, especially if the day has been a long one. Fifteen minutes of oxygen can help sharpen the mind for the task ahead. If the flight has been at moderately high altitude, above 8,000 feet, thirty minutes' worth of oxygen might be a better deal. If you'll analyze the performance you expect from yourself during an approach, the wisdom of using oxygen is quite apparent.

## INTO THE FUNNEL

The airplane is now moving into the funnel. If the pilot remains one step ahead of the airplane, and is confident of the proceedings, things should go well. If the airplane is flown by a pilot with a blank mind, the machine will soon move ahead of the mind and things won't go very well. Catching up once you realize that flying has outrun the thinking is difficult, and the mark of success is in staying ahead. If you do get behind, there should be a plan for that too, even though such a plan can only provide some time to be used with more discipline than was found up to that point. This might mean a missed approach, but a missed one is better than a botched one every time.

Organization puts workload items in the idle time slots, when they can be best handled. It also helps to identify points in the arrival, and to key items to these points. The first point might mark a time by which you'll want to have studied the approach plate and charged the mind with the task that is ahead. Good discipline is to do this right after listening to ATIS or learning of the approach in use. Don't procrastinate—do it.

The next collection of things is an "in range" checklist. This would include final positioning of the fuel selector or selectors if applicable, turning on boost pumps, telling the passengers to be quiet for the remainder of the flight, and every other item from the landing checklist that can be handled well in advance. Some radio preselection can be handled at this time, as applicable. This in-range time might be keyed to being told to contact approach control, or when cleared to descend below 6,000 feet above field elevation in the case of an arrival where no approach-control facility is involved. Time or distance can also be used as a key.

## NEXT

Getting those items out of the way early means that the final landing checklist is minimized, and the items on the final check can be used as helpers in the approach itself. In a retractable, the landing gear is best extended over the point at which the descent to MDA or DH begins. That gives it a dual role. The first is obvious: wheels to land on. The second is to establish the descent. Flaps use varies from airplane to airplane, and the objective should be to use flaps consistently if possible, and to have a plan for flap extension.

If the approach is to a busy airport and the controller makes the usual request to keep the speed up as long as possible, have a procedure for that. Know what the speed will be, and be honest with the controller:

"Zero Three Zero, cleared for the ILS approach, call the tower at the outer marker, keep your speed up as long as possible."

"Roger, and the speed will be 125 knots on final."

That speed happens to be the maximum for a Cardinal RG. It is the top allowable speed with the landing gear extended, and there is no way to fly a good approach at higher speed. As a matter of practice, I fly all approaches in this airplane at about 120 knots to get down and out of the way in minimum time, whether or not another airplane is following along behind. This makes every approach about the same, and the relatively high speed on straight-in approaches, ILS or otherwise, is no problem because the airplane slows rapidly when the flaps are extended (approach flap speed is 130 knots) and power is reduced.

Have some mandatory altitude call-outs. I use 1,000 feet, 500 feet, and 100 feet above MDA or DH, plus the minimum altitude itself. The call to myself at 1,000 feet is both a reminder of MDA or DH plus a reason to verify that the airplane is indeed 1,000 feet above that altitude. Hopefully I would catch any mistake in reading the altimeter at this time. Five hundred feet above is about a one-minute warning. One hundred above is the reminder to be cocked and ready to blast off into the missed approach if the runway isn't in sight after I have descended another 100 feet. The MDA or DH call signifies the time to go away unless the runway is in sight.

## DO IT

Let's run through the thinking and planning that we'll be doing in an approach. I use the Cardinal RG as an airplane of reference because that's what I happen to be flying as this is being written. Adjust as necessary to fit your particular airplane.

"Fifty miles out now. ATIS says it's to be an ILS to Runway 4. Weather is 200 overcast and a half mile visibility, runway visual range to be given by approach control. Altimeter 29.72 and set. Surface wind is calm. Wind aloft has been rather strong southwesterly. That means a decreasing tail wind on the approach. A decreasing tail wind will make the airplane want to trend high on the glideslope while passing through the altitude range where the wind changes from southwesterly to calm.

"Cleared to 5. Contact approach control. I do those things and check that the fuel selector is on 'Both.' That is about all there is to do in this airplane until closer.

"The localizer frequency is 110.3. I set it on number one and identify. The locator at the outer marker is 369; I set it on the ADF and identify. The marker-beacon audio is on. Final approach course is 041, the decision height is 455 feet, 200 feet above the touchdown zone. No fudgee; there's a 501-foot obstruction practically on the final approach course, not far from the middle marker. Missed approach is straight ahead to 2,000. There's a 2,272-foot obstruction about 15 south of the airport.

"Cleared down to 2,000 now. Vector heading is 130 degrees, that's 90 to the localizer. The ADF is 45 off to the left and moving slowly, so I've distance to go before reaching the localizer. leave the speed up at 140 knots in the descent until closer in.

"Getting closer, level at 2. Put number 2 on the localizer

now. (Some would put it in the VOR as a double-check on the outer marker, but I'd rather have them both on the localizer. If they don't agree, then it's time to become suspicious.)

"Left to 090, that's 50 degrees to the localizer. Tower frequency is set in the number 2 communications radio. Flip of the switch will do it when he says to call the tower. Speed is stable on 125 knots so the gear can go down at the marker. The decision height is 455.

"Four from the marker, cleared for the approach, call the tower at the marker. Okay. Runway visual range is 3,000 feet now, well above minimums. No heavy airplane ahead, so no sweat on wake turbulence. Mixture rich. Everything is set. All I have to do now is extend the gear over the marker, call the tower when convenient after that, and fly the ILS.

"Glideslope needle is coming down. There. Gear down, lower the nose as necessary to maintain 120 knots. Add a touch of power, 2 more inches. Marker beeped, ADF reversed. No doubt on position. All settled now, needles crossed, call the tower.

"Okay, cleared to land. The airplane is trending high on the glideslope. Diminishing tail wind. Take off 2 inches of manifold pressure. Be ready to add it back. When the airplane is through the area where the wind shifts and has adjusted to the new situation, that power setting will result in its going low on the glideslope. Decision height is 455, now at 1,455, that's 1,000 feet above DH and descending. Needles crossed, airspeed on 120, that's 700-foot-per-minute rate of sink. Some light turbulence.

"Five hundred above DH, needles not quite crossed. Within two dots, okay to continue. Airspeed is 120 knots, rate of sink is 500 feet per minute, power is back up now, to normal approach value. Air smooth.

"Hundred feet above DH now, prop to flat pitch, cocked for a go-around.

"DH, look up, lights, runway. Do not change anything. Keep flying the localizer and glideslope—they lead to a safe place on the runway. No flaps used on the approach, land without them. Runway is 7,012 feet long, no need for a configuration change so I don't do it.

"Landing was only tolerable, take the first left, cowl flaps open after clearing the runway, call ground control, cleared to the ramp.

"Parked and secured. The planning of that flight has now been completed."

No joke—the planning of a flight is never completed until the airplane is parked. Flying is the mechanical part, planning is the thinking part. Doing it on the gauges requires an equal measure of both, and both continue until the airplane is parked.

# 7

# THE POINTS
# OF STRESS

Every pilot has stress points, some have more than others, and instrument flying tends to involve more of them than other types of flying. To the mind, there's a substantial difference between operating an airplane when you can see mother earth and when you can't.

If you ask a hundred pilots what's hard about IFR flying and what's easy about IFR flying, you might well get a hundred different answers, too. The person with maximum cool, good training, and recent experience will insist that it's all a piece of cake, except of course when cumulonimbus menace the flight path, when ice coats the wings, or when every airport within fuel range is below minimums. The nervous sort who doesn't have much actual experience and who trained in good weather under the hood might break out in a cold sweat at the thought of flying through any cloud, much less a bumpy, frosty, or low one.

## ────────── *LOOKING INWARD* ──────────

It is easy to identify the stressful moments in an instrument flight. Right off the bat, it is common to find tension at the beginning of a flight—especially if there is some doubt about the weather or if the pilot hasn't flown in actual instrument conditions for quite some time.

The careful checking of weather and the deliberate process of deciding that it is okay to start the flight can help on the weather question, but often nothing eliminates all initial queasiness. Somehow we seem to forget from one time to the next that clouds are not solid, and that the airplane will fly through them perfectly well. Instrument flying can seem almost contrary to the laws of nature. The best way to combat this hangup is to recognize that it exists. A lack of recent actual instrument time can also contribute to initial reluctance. A salve to use on this is enough practice to know that you can fly the airplane, plus that deliberate preparation for flight that insures that everything is in its place before launch and that the flight will get off to a smooth start.

The importance of the first minute or two of flight is obvious here, too. If the pilot is properly psyched to settle in with the gauges, taking first things first, it is possible to brush away the doubts and cobwebs rather quickly.

A lot of factors accentuate the first stress point in a flight. The airplane is at its noisiest and the atmosphere is more charged than at almost any other time in a flight. The defying of gravity is much more apparent when leaving the ground than it is when cruising level or descending. The sound and feel of the machine, and of flight, is less familiar at the beginning than later on. Discipline is the key, and with it the initial stress point can even be turned into an advantage. If the stress can be put in perspective and allowed to serve as a mandate for methodical operation and awareness, things will go well.

————————————— **NEXT POINT** —————————————

The next point of stress likely comes with any indication that things are not going exactly according to plan. Anticipate such an event, because there is at least one in virtually every flight. One mark of a good pilot is the ability to accept whatever comes along and move forward with a methodical plan to handle any glitches. A good example can be found in our dealings with the air traffic control system. Perhaps the original clearance is as filed but later in the flight the controllers decide that a more circuitous routing is necessary or more to their liking. This is aggravating, and whereas composure and concentration might have been the norm before, the pilot is likely unhappy and not concentrating on anything properly after the issuance of an undesirable new clearance. Many will argue with the controller.

In any difference of opinion with air traffic control there is nothing wrong with registering a brief protest with a request to return to the original route or whatever you desire when and if possible. Then, to move away from the stress, the only thing to do is move enthusiastically into the task of following the new clearance and replanning the rest of the flight. A new fuel calculation will have to be made, and if the change involves a substantial change in routing, there will be new weather to obtain. Apply yourself to the chores and the stress of the moment will go away.

## DO YOU READ?

Difficulty in communicating can create stressful moments if it isn't handled properly. Pilots (or controllers) sound very calm and matter-of-fact on the first call. If there is no answer, the second call is with a slight sense of urgency in the voice. The third is with more urgency, and you can hear composure crumble in repeated and unanswered calls.

Why be bugged about communications difficulty? There is always a procedure or a plan by which you are expected to continue the flight if communications become impossible for any reason. Have confidence in Plan B and any aberration in Plan A won't be so bothersome. Too, probably 99.9 percent of the communications difficulties are resolved before a pilot has to resort to radio-failure procedures, simply because a high percentage are caused by some error in the cockpit. The squelch might be improperly adjusted, the incorrect frequency might have been selected, a switch in the audio panel might not be in the right position, or the volume might be down.

The first question during time of communications difficulty should be: What have I done wrong? If it is determined that you have not done anything wrong, then study the situation. We fly general aviation airplanes at relatively low altitudes; perhaps the communications site we are trying to reach is simply out of range. Perhaps the controller is busy with something else and can't answer immediately. When the pilot has done everything correctly, these two items are the cause of almost all the rest of communications problems.

Finally, if things remain silent, just remember that communications difficulty in no way affects the laws of aerodynamics. The certificate we carry is a pilot's certificate, not a communicator's certificate. Always concentrate on flying first and handle any communications problem as time allows. Above all, don't let a communications problem cause a level of stress that creates problems in the area of aircraft control or navigation.

## TURBULENCE

Turbulence is a great cause of stress, and we learn early that there is a lot of turbulence in instrument flying, even if we never get within five miles of a thunderstorm.

A primary reason that turbulence induces stress is that it makes us wonder if we have miscalculated. Is this the beginning of a thunderstorm that I somehow missed and the controller did not mention? Will it get worse? Isn't that rain heavy? Will it get heavier? Should I slow the airplane to maneuvering speed? How long will it last? Sure is getting dark outside.

One's gut can tighten and the taste in one's mouth can become quite bad in such a situation. Those are two of the best indicators of stress, so recognize them in such a situation and try to generate a little inner peace. Settle down. Fly the airplane. That's what counts. Flying the airplane gets it through. Worrying about it does not. With the best possible information, there should be nothing there that will really bite. Relax and analyze the situation. Perhaps a lower altitude would be better. Or higher. A few bumps don't signal the end of the world. Hang on, fly, and plan.

Often we make turbulence stressful by fighting the controls. The bumps probably wouldn't be bothersome when we are clear of cloud, but they become of almost consuming interest when we are in cloud. In such a situation I note the airspeed and vertical-speed excursions that are resulting from the turbulence. They are often rather minor, and this tends to classify the level of turbulence and put the mind somewhat at ease.

## THE LEANS

Any pilot who professes never to have had a touch of spatial disorientation, often called the "leans," either hasn't flown much or doesn't recognize a problem that affects all pilots sooner or later.

The leans can come at the oddest times. I recall vividly flying in the Houston area one hazy day and getting such a bad case of the leans that it took intense concentration on the instruments for quite a long while to overcome the malady. I was not really on instruments in the sense that the airplane was continuously in cloud. But with the wings level and the ball in the center, I felt that the wings were not level and the ball was not in the center. I almost *knew* the instruments were telling falsehoods, and I'd have very certainly moved the controls in a manner contrary to that which was proper if I had allowed anything other than the instrument indications to hold sway in my mind.

In this case the trouble was caused by differential lighting, among other things. There was an immense thunderstorm off to my left. It was very dark in that direction, and the very hazy sky was brighter in other directions. Too, I was flying through the tops of little building cumulus of varying heights. I had visual cues but they were misleading, and even as I concentrated on the instruments my peripheral vision was still feeding those confusing visual cues to my mind.

The influence of visual cues can be seen in many other instances. One comes when lying below the level of the tops of building cumulus. There is certainly no natural horizon to use for reference, but there is still always a strong temptation to use the visual cues. You can see, so look. Trouble is, there is vertical motion in the visual cues and you'll almost always climb when trying to fly level in such a situation. The remedy is to fly the gauges.

Sloping cloud tops can also create some wild sensations. There once was a collision between two airliners that came after one pilot perceived a conflict that actually did not exist. The airplanes had the required vertical separation to begin with, but they were on top of a sloping cloud layer, and an evasive maneuver prompted by the illusion led to the tangle.

Another version of the leans can come when you divert from the primary task of flying instruments at the wrong time.

It is bad practice in a turn to try to do anything other than fly the airplane. You shouldn't try to copy a clearance when turning. Trying to read a chart in a turn is bad news. And even communicating when in a turn can be distracting. For one thing, moving your head around can be spatially disorienting; for another, when an airplane is turning it is just that much closer to a lateral upset than when in level flight. Both are good reasons to concentrate on the gauges when in a turn. If you don't, you can have stressful moments.

Picture yourself in a 20-degree banked turn when the controller calls with clearance to the Bltzf intersection. Write it down. To the chart for location of the intersection. The wheel gets moved a little bit. The bank steepens, the airspeed increases. You perceive that something is awry, either through a change in the sound level or through a glance at the instruments. To restore full vision to the panel you have to move your head, which in itself is a promoter of disorientation. Now comes the crucial and immediate need to interpret the instruments correctly and take action. A very stressful moment, as well as likely time for the leans. There is a 50 percent chance of moving the wheel in an incorrect direction, and in this situation you might well come up with the wrong answer.

Again, any diversion from the instruments when turning is bad. It is much better to ask the controller to stand by or to request a repeat of the clearance after the turn is completed.

Even when you are straight and level, certain techniques need to be used to minimize problems when writing things down or consulting charts. The primary point is to have the airplane trimmed and to release the control wheel when looking at something besides the panel. That precludes the inevitable inadvertent slight movement of the controls that comes when your mind is in one place and your hands are on something else.

## CONFUSION

Stress also enters the picture whenever we have a feeling of being behind the workload curve. In planning an arrival, doing things in advance is for the purpose of avoiding moments of ignorance or indecision, things that lead to stress. If you forget to listen to the ATIS information and the controller tells you to let him know when you have it, the flight is punctuated at a point where you didn't want or anticipate a pause for information. If you don't study the plate in advance and have to find the decision height on the chart after passing the marker inbound, that diversion will move the thought process from keeping the needles crossed to interpretation of the chart, and the result can only be harmful.

## ANTICIPATE

When moving into the arrival phase of a flight, try to avoid stress by anticipating any special information needs. For example, when flying light airplanes we don't have to give a lot of thought to wind and runway lengths for ILS approaches, but in nonprecision approaches to small airports these can become critical items. I recall two approaches that illustrate this point very clearly.

On the first VOR approach the straight-in minimums were quite a bit lower than the circling minimums. As I moved toward the MDA it was quite apparent that it would have to be a straight-in approach. I continued, and was about to give up on even a straight-in when the end of the runway became visible not too far ahead. I was at MDA for the straight-in and below the circling MDA. The decision had to be made in an instant: straight-in or missed approach? My decision was that the airplane was in a position from which a normal landing could be made. I chopped the power, extended full flaps, and headed for the runway. Trouble was, I didn't *know* either the wind or the runway length. I could have easily had both—the runway length by looking at the chart and the wind by asking—but I had neither, and was conducting the important final portion of my flight based on eyeball and ignorance. The runway was probably around 3,000 feet long, I thought, and the southerly flow wasn't too strong, I hoped. I landed and stopped, thankful for good brakes. There were a few stressful moments along the way.

A similar approach was later shot to home base. The runway appeared a little to the right and ahead at the straight-in MDA. I wanted to circle to land into the wind, which I knew was from the south at 10 knots, but there was no way. The ceiling was just too low. Again, it was a straight-in or a missed approach. The difference was that my mind was filled with information this time. I remembered that the pilot's operating handbook gives a number for downwind landings with up to 10 knots, and I had once figured that it would take at least 1,200 feet to stop with a 10-knot downwind after a normal touchdown. I knew the runway length, and I knew there would be 1,700 feet left if I landed just before reaching the taxiway. That left 500 to spare. So I continued and landed. I used just as much brake in stopping as on the previous approach, but there was less stress, or doubt, or whatever you want to call it, because I was operating with a set of known quantities that all suggested success.

## SINGLE PILOT

I have heard some general aviation instrument pilots say that it is thought dumb to fly with just one pilot in the airplane in areas with a lot of IFR traffic. Or they at least insist on a good operative autopilot when flying to such a place. This feeling is no doubt related to some feeling of stress that has been encountered when going to and from a big terminal.

In reality, there is no valid reason why any pilot should find it stressful to operate a light airplane IFR in the busiest terminal area in the world. All the rules are the same, and the only true difference in the operation is in the number of other aircraft the controller is talking to at the same time. If there are enough, the informal and chatty patter we find at some smaller terminals goes away, replaced by very crisp and businesslike talk. If just hearing that induces stress, the pilot probably needs more help than is available from another pilot or an autopilot.

In light airplanes we have the advantage of flying rather

slowly through procedures that are designed for faster airplanes at the busy terminals. So instead of letting the busy
spots bug you, just remember that while the Gates Learjet
pilot might indeed have a copilot to help with the chores, a
Cherokee pilot has twice as much time in which to do them.

The amount of work at hand, or the amount of information coming in, can have a definite bearing on stress, too. in
fact, studies have shown that overloading the brain can all
but cause it to cease functioning. Given too many things to
think about, or sort out, a pilot won't be able to make sense
of anything. Brain management at such a time is critical.
You have to pause, regroup, and all but start over at the
beginning. Air traffic control personnel use the phrase "stand
by" rather freely; a pilot should do the same. If a clearance is
incomprehensible, it could well be because there was enough
there to paralyze your thinking process. This is often your
fault, for going in cold without studying charts in advance to
know the lay of the airways and the names of intersections
and Vortacs likely to be used in clearances, but while it is
happening the drill is to improve the situation, not to lay
blame. If the brain has a severe overload from some new instruction, it's certainly in order to ask for a pause with "Stand
by," and then request that the new clearance be read more
slowly.

## EDUCATIONAL

We can learn a great deal from our moments of stress.
They should be remembered for study and future reference,

because they tell us a lot about our weaknesses in creating bad situations and our ability to recover from a burst of stupidity by handling the situation. Whenever there is a bad moment, spend some time reflecting on it later, in a quieter time. Dissect the event and your response, and think both of ways to avoid such situations and of ways to handle them with aplomb when avoidance procedures are unsuccessful. Don't ever pass off a bad moment with the thought that it couldn't have been *too* bad because the flight did end in a successful landing. There is a very fine line between an airplane parked on the line and one wadded in a heap, and any event that is the least suggestive of the latter must be considered serious.

Some of our most stressful times come in the special IFR moments when we become involved with thunderstorms or ice. Night IFR is a different league, too, and all three of these are worthy of their separate chapters, which follow.

# THUNDERSTORMS!

By some estimates there are as many as forty thousand thunderstorms a day in the world. Pair this with the fact that forecasters are likely to include the chance of thunderstorms if conditions are the least ripe, and you can see why general aviation pilots spend so much time thinking about thunderstorms.

As frequently as they occur, thunderstorms do not devour airplanes at an unreasonable rate—we don't lose many general aviation IFR flights in them annually—but that knowledge is of precious little comfort when we're being banged around and hosed down in what seems an unending maelstrom. The things are bad, very bad, and the advice is always simple: Don't fly into a thunderstorm. Preachy, and it often seems easier said than done.

Most of what is written stresses avoidance, and as you listen to the radio chatter in an area where storms are active you hear pilots trying to heed that advice, pleading for vectors around activity, or at least information; a lot of users pay big money for airborne weather radar for use in steering clear of cells. As radar becomes readily available for single-engine airplanes, it's bound to become even more popular.

Radar does see precipitation, and it is true that there is a relationship between precipitation and turbulence. But it is also true that severe thunderstorm-related turbulence can be found away from precipitation. The accident record clearly tells us that there is more to staying intact than just avoiding the areas of precipitation in thunderstorms. It is quite a challenge, because if our airplanes are to be useful, we must operate in thunderstorm areas but not in thunderstorms. We must learn how much we can tickle the tiger.

Experience suggests that we back up what we see as we fly, plus any radar information (ground or airborne), with a strong basic understanding of thunderstorms. Without the basics, all the help in the world might not be enough to see a pilot through an area of activity and to the sunshine on the other side. So let's look first at the storm.

## THREE REQUIREMENTS

There are three basic requirements for thunderstorm formation: unstable air, lifting action, and a high moisture content in the atmosphere.

Stability, or instability in this case, refers to the atmosphere's resistance to vertical motion. In very general terms, if the rate at which air cools with altitude is more rapid than normal, then the air tends to be unstable. If, in unstable conditions, a particle of air is given a little upward shove it will tend to keep going up, and even to accelerate. This makes turbulence, and if there is moisture around it gathers it up and uses it to make clouds.

A hot summer afternoon provides the best example of instability. As the surface heats, the warm air rises. The rising air cools, and puffy cumulus clouds form at the level where the rising air cools to the dewpoint. Lifting continues where the clouds form, and they billow on upward. If there is enough of everything, nature can get it all together and create a thunderstorm.

On the other hand, in stable conditions a piece of air that gets a little vertical nudge tends to just move up an amount about equal to the nudge and stop. It can even settle back to where it was. So no vertical development is possible. If there is warm air above cool air it's called an inversion and things are really stable.

The lifting action that gets things started in unstable conditions can come from heating by the sun, from wind flow over mountains, or from the collision of air going in different directions with resultant vertical motions.

Moisture? There is always some in the air—sometimes enough to make thunderstorms when all the other ingredients are present, and sometimes not enough to make anything.

## BUILD ONE

Once all the ingredients are present, how does nature construct a storm?

A cumulus cloud is the basis for a thunderstorm. The action is upward in the cumulus. In unstable air the vertical currents accelerate as the warm moist air of the cloud rises into the colder air above. The cloud feeds itself, attracting moisture from the surrounding atmosphere. In this cumulus stage, the updrafts extend from near the ground to a bit above the top of the cloud. The biggest upward push is toward the top of the cumulus cloud, where the vertical velocity can be as high as 50 feet per second.

If there is enough lift and if the air is unstable enough, the cloud top goes on up through the freezing level. At some point in the development the old saw about everything that goes up must come down takes over, and the moisture starts down—as rain or as hail if the moisture particles are held aloft long enough to freeze.

The precipitation usually starts within ten or fifteen minutes after the top of the cumulus builds through the freezing level. If there is not enough lifting action and instability for the cloud top to make it through the freezing level, there won't be a thunderstorm. This is why, even in the summer, the freezing level is a useful bit of information. If you know the freezing level as you are moving along through a thicket of building cumulus, you will also know the point past which the tops must go to begin to qualify for admission into the thunderstorm club.

When the moisture starts down as precipitation it brings air with it, forming a downdraft in what has just become a mature thunderstorm. The updrafts remain active around the outside of the storm at this stage.

As the downdraft approaches the surface it is deflected outward, and strong surface winds can result. The strongest wind will be in front of the storm, on the side toward which the storm is moving. The velocity of this wind is said to be the sum of the downdraft velocity and the forward speed of the storm over the ground.

## THE CLIMAX

Early in the mature stage of a storm—right after the downdraft and precipitation make their move—the surrounding updrafts tend to increase and reach a climax. This is the meanest moment of the life of a storm; it is all downhill from here on. The updrafts begin to subside as the moisture falls out of a cloud and brings cool air down. Soon the updrafts are pretty well gone and it is mostly downdraft. Things are relatively tame then.

The life cycle of an individual storm cell is from twenty to ninety minutes, but if conditions are ripe, cells can form in lines or clusters; there can be gaggles of cells in various stages of development, maturity, and dissipation at any given time. What might look like one thunderstorm could really be several separate cells. The maturity of the group might run for several hours. Or, if the lifting, instability, and moisture supply are continuous, as with a squall line ahead of a cold front, the turbulent collection of clouds can exist until one of the factors is modified.

## DRAFTS

Updrafts in a mature storm tend to increase in velocity through the lower two-thirds of the storm, with maximum draft velocities found between 14,000 and 20,000 feet. They often reach 60 to 70 feet per second, or more. Updraft strength is greater than that of downdrafts, but these also accelerate; the taller the storm, the stronger the downdraft. The 40-knot gust ahead of a storm converts to 67 feet per second, or 4,020 feet per minute, for some idea of strong downdraft strength. You would have to subtract some from that for the forward motion of the storm, but even 40 feet per second is quite strong. That is 2,400 feet per minute, which is well in excess of the rate of climb of most airplanes.

The air within a downdraft isn't so terribly turbulent. In fact, research has shown that the least turbulence in a storm might be found in the area of heaviest rain, in the downdraft. That is, of course, relative. It might still be plenty bumpy.

What really hurts is moving from updraft to downdraft and through all the shears and eddies that develop when air rushing upward is next-door neighbor to air rushing downward. It is easy to visualize the tremendous turbulence that would develop in such a situation.

## EXPERIMENT

It is also easy to estimate the effect of a thunderstorm on our airplane's ability to maintain altitude. Consider a garden-variety storm with maximum updrafts of 25 feet per second and downdrafts of 18 feet per second, just for example. The updraft will be equivalent to 1,500 feet per minute and the downdraft equal to 1,080 feet per minute. Now, flying a light airplane at maneuvering speed, note the available rate of climb at full power and the available rate of descent when power is off. If the rate of climb is less than 1,080 feet per minute and the rate of descent is less than 1,500 feet per minute, then maintaining altitude is unlikely even with gross use of power. There are other factors that might affect rates of ascent and descent in a storm, but the example is useful for illustration. And remember, that is not too much of a thunderstorm.

Another useful mental exercise is to relate the turbulence from wind flow over rough terrain to that in a thunderstorm. Thirty knots is equal to 50 feet per second; a 30-knot wind over rough terrain can do some pretty enthusiastic things to an airplane flying downwind of the terrain. And consider that the force of those up- and downdrafts is mitigated by distance.

The point that must be understood is that it is extremely turbulent in and near any active thunderstorm cell, and the up- and downdrafts are likely to be more than an airplane can handle, even if the pilot puts maximum effort into maintaining altitude. That's why the word on flying technique is always to maintain attitude and not to worry over altitude variations. Theoretically the altitude might average out during passage through the storm.

## SHEAR

As we consider the air rushing in around the sides to form the updraft, and the downdraft in the center, we come to a logical reason why the worst turbulence is often found around instead of in storms. The downdraft *must* fan out as it nears the ground. The cold air of the downdraft pushes under the warm air around the storm, and as the storm continues gathering moisture from the surrounding airspace, air rushing in to feed the updraft is above the downdraft air. There is great turbulence where the two rub against each other—and this can be found quite a distance away from the precipitation, in what is commonly called wind shear. The outer limit of severe shear should be at or above the point on the surface reached by the first gust from the storm, but turbulence related to air moving into the storm can be found farther away than that. Many storms give us a picture of the rolling and tumbling of air that takes place in the shear area between the updraft and downdraft. The illustration is in the roll cloud that forms a thousand or more feet above the ground, ahead of the storm cloud itself.

In visualizing the turbulence caused by the interaction between the updrafts going in and the downdrafts coming out, also visualize how there can be a monumental amount of disturbance in the air between storm cells when conditions are ripe for thunderstorm development.

---------------------------- *LESSON* ----------------------------

Let's take what we have considered to this point and relate it to an air-carrier accident. It is very pertinent to general aviation flying because the airplane was only 4,000 feet above the ground as the accident started happening, and that is an altitude that we frequently use.

A line of thunderstorms lay between the point of departure and the destination, with the storms moving toward the point of departure. The pilot was aware of the line, and of the fact that it was an enthusiastic example of thunderstorm weather.

The flight requested a low altitude for penetration, and as the plane neared the weather, the crew sought the controller's opinion on the situation. The controller told them that the line of weather appeared solid. The pilot also talked with another air-carrier flight in the area, which was in the process of penetrating the line from the other direction, at a much higher altitude.

After considering the information gathered, and with the use of airborne radar and visual observations (it was dark, but there was a full moon in clear sky ahead of the line, as well as considerable lightning in the storms), the pilot requested a deviation from course. One can only assume that the deviation was to head for the most favorable point for penetration, based on the information at hand.

The airplane never reached the line of thunderstorms. According to information from traffic control radar, it broke up five or more miles from the nearest observable echo at the time of the accident. There was a roll cloud in the area, and the airplane apparently just reached the area of this cloud when it flew into catastrophic turbulence. Ground witnesses reported that some rain fell after the roll cloud passed, but no heavy rain was reported until forty-five minutes later. Two funnel clouds were observed half a mile from the accident site approximately eight minutes after the accident. (Reports of funnel clouds from the general public are often related more to the churning in and near a roll cloud than to an actual tornado. However, conditions this evening were ideal for tornado formation.)

Two other air-carrier flights came through the squall line from the other direction at approximately the time of the accident. Crew statements and flight recorder readouts both indicated that the heaviest turbulence was encountered during a short period *after* passing through the precipitation. One of these flights flew through at about 2,000 feet above the ground; the other was more than 15,000 feet above the ground.

The airplane that was lost simply came to the wrong place at the wrong time. The storms were very tall with strong downdrafts, and there was continuous cell generation within the line of storms, because there was plenty of air rushing in to feed the updrafts.

The airplane was flying at the proper turbulent-air-penetration speed as it approached the line of storms, but in flying into the shear area the aircraft experienced a momentary and substantial increase in airspeed. This could have been caused by a horizontal gust, or the increase in airspeed could have come as the airplane flew from an area where it had a very strong tailwind—it was flying toward the storm, and air at its level was being drawn into the storm—into an area with no tailwind. Airspeed will increase momentarily in that situation.

Whatever, the next ingredient was another gust. According to the accident report, this one probably was an angled gust and very strong. Visualize it as a stream of air, like a stream of water from a high-pressure hose, rushing upward at a 45-degree angle. The airplane flew into this with the airspeed still well above the turbulent-air-penetration speed—a situation caused by the gust encounter or abrupt tailwind decrease of a moment before—and the loads imposed on the structure as the airplane responded to this last gust were in excess of the limit load factor. The vertical fin and tail structure broke off almost simultaneously.

Again, this happened five or more miles from the closest weather return shown on the controller's scope. The airplane was at the proper penetration speed to begin with, and the low-altitude penetration conformed with a long-held theory that lower is better around thunderstorms. The government publication *Aviation Weather* had put it thus: "The softest altitude in a thunderstorm cloud is usually between 4,000 and 6,000 feet" above the ground.

With this example in mind, let's examine the considerations of speed and altitude.

## HOW FAST

In this accident case, the speed was correct to begin, but the movement of air ahead of the storm outfoxed the crew, caused an airspeed increase, and then took what might seem unfair advantage of the aircraft with another gust.

I think that we have to recognize that such an airspeed increase is to be expected when approaching a thunderstorm, especially when flying toward the direction from which the storm is moving. Every factor in the construction of the storm works toward an increase in airspeed as the turbulent part of the weather is reached. The decreasing tail wind experienced as we fly from the area where air is rushing toward the storm to feed the updraft and into the shear between the updraft and downdraft is not to be dismissed lightly. For example, if the true airspeed is 200 knots and the groundspeed 240 knots because of a 40-knot tailwind, the groundspeed will tend to remain constant at 240 knots momentarily as the airplane enters the area of no tailwind. Quite a spike in airspeed will be experienced as a result.

Or, just considering the updraft itself, flying into air that is rushing upward will cause both an ascent and an increase in airspeed. Add this to the fact that the airplane is flying with a decreasing tailwind—the air is turning upward as it feeds the storm, and we are flying from an area where the flow was more parallel to the ground, affecting the airplane as a tailwind, and into an area where the tailwind will be lost and the updraft will be experienced—and you can see another reason why the airspeed increases.

As we fly toward a thunderstorm, this increasing-airspeed phenomenon makes the airplane seem almost like a sheep being nudged by a sheepdog. The nose seems to want to drop slightly as the airspeed increases. It should be noted that the decreasing-tailwind phenomenon is related more to low-altitude operations where the downdraft is starting to fan out over the surface of the earth. Higher, above 3,000 or 4,000 feet, all the action is more nearly vertical.

There is a dual moral. First, don't be in the position of trying to descend when near a thunderstorm cell. This can only compound the airspeed-control problem that might be encountered. If you want to be at a low altitude, get there well before reaching any area of questionable weather. Second, consider flying the airplane at a value somewhat *below* the turbulent-air-penetration speed if you suspect that the inflow/outflow effects of a storm might affect the airspeed. Flying at a speed below that recommended does bring added risk of stall if a strong vertical gust is encountered, but when we are talking about doing something that we really shouldn't do, such as flying near a storm, it becomes a matter of the lesser of risks.

The effect on airspeed reverses itself as you fly out of a thunderstorm, and the result can be a very bad sinking spell. When flying out of a storm at low altitude the airplane is likely to experience an increasing tailwind component as the downdraft turns parallel to the ground and a rapidly increasing tailwind causes the airspeed to decrease. It can feel almost as if someone pulled the chair out from under you.

All this updraft-and-downdraft business, and the interaction between the two, should offer a pretty plain illustration of why it's next to impossible to maintain altitude in a mature storm, and why the airspeed will fluctuate and the turbulence will be rather wild if we attempt to pass near or through a storm.

## ALTITUDE

There have always been strong feelings about the best altitude to fly when working in thunderstorm areas. "The lower the better" is usually the consensus (except when you can top all clouds by a comfortable margin, but that is reserved for jets and they can't always do it), but there are challenges to the theory. In thunderstorm-research flying, reports have been made suggesting equal turbulence at all levels. But the research flying is in high-performance military jets, and they don't penetrate storms below 6,000 feet above ground level. They have never really tested the 4,000-to-6,000-foot theory, or the feeling of many (the author included) that lower than 4,000 is even better in light airplanes. This is left to study and the imagination.

Again considering the construction of the storm, the lowest possible altitude would tend to add risk of wind-shear encounter, but it would tend to minimize the risk of encountering a catastrophically strong vertical gust except in an extreme case such as encountered by the air carrier in the accident just related. When choosing between the devil and the deep blue sea, between wind shear and strong vertical gusts, wind shear seems the better one to accept in a light airplane. The light airplane's airspeed will be less affected by wind shear, because the airplane adjusts more quickly to a "new" situation than does a heavy airplane with more mass momentum. Too, we can be willing to use power grossly in a light airplane, where you just don't do it that way in a jet. The advantage is reversed when it comes to vertical-gust encounters, because an airplane's reaction to a vertical gust is more in proportion to wing loading than anything else, and larger airplanes have higher wing loading. The lowest possible altitude seems to me the best in a light airplane; higher is probably better in a jet.

When flying at low altitude, the downdraft must be considered. It is true that a downdraft will not take an airplane to the ground, but it is equally true that a downdraft can put an airplane in a position where a collision with the ground is inevitable. The lower-is-better theory thus must be modified when the terrain in a storm area is anything other than flat. And even then a strong enough storm could probably get you because at some point that downdraft will result in a rapidly increasing tailwind for the airplane and there might not be enough altitude to handle that problem. But, again, the light airplane handles wind shear better than the heavy airplane.

Several other things can be used to bolster the lower-is-better theory. One relates to the imbedded thunderstorms. These are often associated with warm fronts, where we find cold air near the surface and warm air aloft. (The height of the base of the warm air depends on the distance from the surface position of the front.) The meteorological fact of life is that the cold air is stable, with the instability and the thunderstorm bases aloft, on the slope of the warm front. Often a general aviation pilot will give a brave report on weathering a storm, and say "It wasn't so bad" in true ace fashion, when actually the pilot only flew *below* a storm that had a base at a high level. This is an entirely different situation from when the air is unstable from the ground up, and there is no question that low altitudes can produce good rides when the bases of the storms are high.

--------------------------------- *SEE* ---------------------------------

Another thing in favor of flying low is found in the possibility of staying in visual meteorological conditions so that you can eyeball the way ahead and avoid the areas of heavy precipitation by a good margin. If there is general rainfall, some success can be found in always flying toward the areas that are lighter in appearance and avoiding the areas that are darker in appearance. Always do this though, with the knowledge that the moment might come when all areas are dark in appearance. One old rule of thumb that has served me reasonably well over the years suggests that you can fly through at low altitude only if you can see through the rain and to the other side. This hasn't precluded rough rides, but it has worked.

Another technique that can offer guidance in certain situations involves looking 30 or 40 degrees above where the horizon would be if you could see it, noting light and dark areas at that level, and flying toward the light areas.

Whatever, there is no doubt that the severity of a storm has at least as much to do with survival as does any operating procedure. Too, the mechanism that spawns the storm—a warm front, cold front, heating, or lifting along a ridge line—has a lot to do with best operating altitudes, and it is an absolute certainty that severe storms offer only various degrees of impossibility.

─────── **ANOTHER OLD WIVES' TALE** ───────

There's another thunderstorm parable that must be examined. It suggests that, once in, the best way out is straight ahead. There is another air-carrier accident to use in studying this. Not picking on airliners, mind you, but the lessons are so much better when there are flight recorders and cockpit voice-recorder tapes to study.

This aircraft, a turboprop Lockheed Electra, was being operated at high altitude. Upon inspection of a line of storms 60 miles ahead, to the northwest, the crew requested descent to a lower altitude and a deviation to the west. The airplane was cleared down to 15,000 feet. The aircraft was later cleared to 14,000 feet and then down to 5,000 feet, but it was lost before it reached 5,000 feet in a normal descent.

The line of storms was quite strong, and this crew headed toward an area that had been avoided by other aircraft and that did not appear a good penetration spot to the air traffic controller, based on his radar picture.

The cockpit voice-recorder tape suggests that the pilot felt they could go under the line of storms, presumably at 5,000 feet, and the following excerpt from the National Transportation Safety Board report on the accident is worth some study:

> At 1641:07, the captain made another announcement to the passengers advising them that he was turning on the "seat belt" and "no smoking" signs "just in the event it's a little choppy in the area." He stated that his radar was working and he was going to be able to "go well under and to the west of all the thundershowers, but they will be visible to you on the right.

. . ." At 1641:42, the captain said (to the other crewmembers) "I guess I can go under." At 1644:16, the captain instructed the flight engineer to turn on the engine heat temporarily, "at least on number one, till we get about twelve degrees or a clear area." At approximately 1646:30, the captain instructed the first officer to ask the controller if he had any reports of hail, which the first officer did at 1646:32. The controller replied "No, you're the closest one that's ever come to it yet. . . . I haven't been able to, anybody to, well I haven't tried really to get anybody to go through it, they've all deviated around to the east." Following this transmission, the captain advised the first officer, "No, don't talk to him too much. I'm hearing his conversation on this. He's trying to get us to admit (garbled words) big mistake coming through here." The first officer stated shortly after that, ". . . it looks worse to me over there." This statement was followed by the sound of the landing gear warning horn and the statement of the captain "Let it ring." The captain then said, at 1647:20, "Let's make a one eighty" and the first officer requested permission to make the turn, from the controller, 3 seconds later. The turn was approved "right or left" at 1647:26.5. At 1647:29, a sound similar to hail or heavy rain was recorded and, at 1647:30.5, the first officer transmitted "three fifty-two." (That was the flight number.) One-half second later the captain said, "let me know when we come back around there to reverse heading for rollout." There was no recorded reply to that instruction. At 1647:35.2, the first officer said "three forty," and immediately afterward the sound of the landing gear warning horn was heard and the captain said "Right." At 1647:41.3, the sound of a fire warning bell was heard and continued until the end of the recording. At 1647:41.9 a sound appeared that was described as being similar to breakup noise.

## THE BUMPS

The flight recorder records G forces, and the readout was without remarkable excursions until only about twenty seconds before the captain ordered the 180-degree turn. At that time there was an increase of both frequency and amplitude of excursions, but there was no extraordinarily high G force until the aircraft broke up. Other flight-recorder data revealed that the airspeed increased rapidly from 206 knots beginning at about the time the turn commenced, and reached 360 knots in about eight seconds. Apparently the first officer's call of "three forty" was in relation to indicated airspeed.

The readout indicated that the airplane was initially rolled into a 24-degree bank, which was maintained for ten seconds. Then the bank increased to 66 degrees, and then to an even steeper bank as the airplane entered the dive in which it failed.

Another quote from the National Transportation Safety Board report is interesting: "After the penetration of the storm had been initiated, the decision to reverse course was not in keeping with recommended company procedures for operation in areas of turbulence. Normally, once in an area of turbulence, the crew is expected to maintain the attitude of the aircraft as nearly straight and level as possible and maneuvering is kept to a minimum until the turbulent area is cleared."

There is nothing in the flight-recorder data to indicate an encounter with severe turbulence. Rather, the pilot lost control in the turn, and the breakup was probably related to the airspeed buildup after the loss of control. Why did he lose control of the airplane? One item that I think must be strongly considered is related to the fact that noise draws the eye. In this case, there was enough noise from precipitation to be heard on the cockpit voice recorder. The streaming of water over the windshield also draws the eye, and if there was suspicion of hail, I guess the temptation would be to look to see if indeed it was there. That this might have diverted the pilot's attention away from the flight instruments is not offered as any explanation for the accident, but it would be a consideration in turning around in similar circumstances. I know from experience that it takes strong discipline to put 100 percent thought and effort into such a turn.

The corollary that you shouldn't turn at all once in a storm seems valid in this accident. The storm tends to be meanest in the direction in which it is going, and once you are into the heavy precipitation the worst area of turbulence could be behind. This aircraft was in the heavy precipitation when it started to turn. If a turn was started soon after entering the precipitation, within thirty seconds, it would be reasonable to assume that it would take a minimum of two minutes to make the 180 and be back out of the weather—thirty seconds in, a minute for the turn, and thirty seconds back out. In going straight ahead for this length of time, the aircraft would have flown more than six miles and would have been close to the back side of the weather.

Turning certainly increases the risk of loss of control, too. When the wings are level, the airplane is as far away from a lateral upset (which is inevitably followed by a longitudinal upset) as possible. When it is banked, the airplane is just that many degrees closer to a lateral upset.

The 180-degree turn is the perfect way to deal with thunderstorms *only* if the turn is started well before the storm is penetrated. Once in, the best way out might be straight ahead. For my nickel, exceptions might be found after entering an area of weather from the direction *from* which it is moving, from the back side of the line for example, or after entering what has been described as a large area of imbedded activity.

The two accidents related almost seem to have a "can and you can't, will and you won't, damned if you do, damned if you don't" ring to them. In the first case, the jet, it almost seems that only the act of not flying would have offered salvation. In the second, there are more alternatives, but it is still one of those "there but for the grace of God go I" situations. However, both were in severe weather, both weather situations were obvious to the eye and radar, and avoidance would have been simple in a light airplane. If we have and follow good rules, such events can be relegated to the category of the other pilot's problems.

If there is a common thread that teaches a lesson it is in relation to airspeed. Note that in both cases the airplane broke *after* the airspeed strayed from the proper value for turbulent-air penetration. As we poke around in weather let's then recognize airspeed control as a primary task if we are to avoid being devoured by a thunderstorm, or by any other form of turbulence. In most general aviation thunderstorm accidents it is possible to determine that the overload failure came not because of turbulence but because of pilot-induced loads. That generally means that the pilot lost control of the airplane, the airspeed increased, and the pilot pulled it apart in a recovery attempt. There are a few that involve airframe failure caused by an initial turbulence encounter, and in these the airspeed *had to be too high*, because an airplane simply will not break if the airspeed is at or below the correct value. All of which is one way of saying that if the pilot doesn't understand the importance of keeping the speed on a proper value in turbulence, the airplane does. And once something breaks in reflecting the airplane's understanding, there's no way to put it back together.

---

## RADAR

---

Both the airliners had airborne weather radar, and in both cases there was discussion with the controller about the depiction of weather on the traffic control radar. There was quite complete information on the weather, but radar is not magic. Above all, radar is not a device that can be successfully used in penetrating areas of severe weather. The guideline for radar use is to stay 5 miles away from the precipitation return of garden-variety storms and 20 miles away from the precipitation return when severe storms are forecast. Those guidelines are based on experiences such as the two accidents related and on very sound meterological information. When a pilot uses airborne radar to cut it closer than those guidelines, the risk increases drastically.

For those of us who don't have airborne radar, there's only eyeball information plus what we can learn from quizzing the air traffic controller about his radar depiction of weather. And while the information we gain from the controller is far better than nothing, anyone who uses it as something other than broad and general information had best be pretty sharp at flying instruments in turbulence.

Traffic radar is designed to show airplanes, not weather, and the computerized radar systems incorporate circuitry that processes the radar return and displays only the weather deemed to be in excess of a certain precipitation rate, based on radar reflectivity. Back in the good old days, before computerized radar, some controllers could and would vector aircraft in areas of weather, taking them through the lightest areas, and this often worked well. Many IFR pilots have had some magnificent rides through fearful-looking skies with the traffic controller as tour guide.

The vectoring is less precise in weather areas with the computerized radar. Controllers, too, often seem reluctant to vector, because it is above and beyond the call of duty, and if a controller provides the service and the airplane doesn't make it, that is bad. The last vestiges of the good old days are found in terminal areas, where radar systems depict weather better and where controllers often seem more willing to help on weather.

## PROCEDURE

We do need a procedure to use if we are to get information from the traffic controller on a uniform basis. I have found the most effective way to be through a series of questions. Even the most reluctant controller will help if properly quizzed.

"Is there weather along my route of flight?"

"Affirmative."

"Well, would a deviation up along the north airway avoid the weather?"

"Negative."

"Okay, how about a deviation down around by the south airway, would that help?"

"Affirmative."

In other words, even in the most extreme case you can often find the best path with yes/no questions. Most controllers respond with better information, and most will give general information on areas of weather at the first pop of your question, but be prepared for the reluctant one.

Do be wary of the effect of time on radar information. As we noted, cells generate and dissipate rapidly, and an area of no return on radar fifteen minutes ago might be filling in now. If conditions seem ripe, keep asking.

—————————— *EXAMPLE* ——————————

I had a rather classic illustration of fast changes in my airplane as I flew along one day. The synopsis had positioned a low to the west, and the forecasts all called for quite good conditions. None included the possibility of thunderstorms, and as I checked weather it appeared that there would be none. I looked at the weather radar with my own eyes; it was on the 250-mile range, and there was nary a bright spot to be seen.

After takeoff I could quickly tell that all wasn't well. At

6,000 feet I was in and out of the tops of cumulus, the tops were building, and it was generally an uncomfortable situation in appearance. I listened for static on low frequency, on the ADF, and sure enough, there was a little there. (An ADF is indispensable for that. It doesn't tell you where storms are, but it tells you that there are storms somewhere so that you can get to work and locate them.)

I then started a quiz program with the center controller, who allowed that weather was beginning to appear on his scope a bit farther up the line. He suggested a slight deviation to the south, and added that I'd be worked by an approach-control facility when nearer the weather and perhaps they could be of specific assistance.

As the miles slipped by, it was clear that there was something ahead that required both care and thought. Even though nothing had shown on the chart, it was quite apparently a warm frontal situation, so I requested a lower altitude. This was approved, and in talking with the approach controller about weather I found that he wasn't doing very well on vectors. He said that he had just vectored one airplane through what appeared the best area and the pilot had reported heavy rain, lightning, and turbulence. That airplane was at higher altitude, though, and I felt that going on toward what appeared to the controller to be the lightest area was a reasonable proposition at a lower altitude.

When the weather was a few miles ahead, the controller made it plain that there was no "clear" area. He wanted me to understand that he was doing his best but there were no guarantees. I knew that, and went into the area with the airplane a few knots under the turbulent-air-penetration speed. I basically flew toward the lightest spot in appearance, which also looked best to the controller. The ride through was not bad at all, but there was quite heavy rain and some lightning for a minute or so. Then I flew out on the other side.

The flying was done not on the basis of the controller vectoring me; his only word was that a heading of 330 would take me toward the lightest area. What I could see tended to verify this. I felt that my decision to continue was sound enough. I would not have gone through if the activity had been related to a strong cold front or low-pressure center, or if severe weather had been forecast. And in rationalizing away the 5-mile rule on clearance from cells, I flew on the basis of conducting the operation below rather weak storms that had bases at a rather high level.

## HOW TOUGH?

When we start poking around in weather, there's always some question about airframe strength. How strong are the airplanes? The answer is that they are strong enough if operated at the correct airspeed. Certainly there are impossible turbulence situations, as noted in one of the air-carrier accidents, but I think that we can fly general aviation airplanes with confidence in their ability to handle most turbulence *if* they are flown at or below the turbulent-air-penetration speed. And I like to think that because I work very hard at avoiding the penetration of *any* cell, the ones that I might accidentally penetrate should be of a milder variety and will be manageable if only I stick with the task of keeping the wings level and the airspeed at or slightly below the prescribed value.

## PRINCIPLES

There are some meteorological facts that we can store in mind to help in planning storm avoidance.

Storm cells tend to move with middle-level winds, which are generally from the southwest in the U.S. when conditions are ripe for storms. An area of storms might move from the northwest, as in the case of a squall line ahead of a front, but the individual cells are likely moving from the southwest within the area of storms.

New cell generation in a cluster of storms tends to be on the side toward the low-level wind flow. The low-level wind is usually from the south or the southeast, so watch out for cell generation in this quadrant of a cluster and stay well clear. If there is to be hail, it is more likely in front of the storm, in the direction toward which it is moving. If there is a large high-level overhang of cloud off the top of the storm, don't fly under it, as there could be hail there.

Strong upper-level winds, jet streams, are a major contributor to the development of severe storm activity. To simplify completely, the thing to watch for is a low-pressure center on the 500-millibar chart (approximately the 18,000-foot level), because when one of these is positioned to the west of your route of flight, cold air aloft is being drawn down around the south side of the low and then moved up over the warm air east of the low. When the jet stream takes such a trip, it creates ideal conditions for thunderstorm activity. Because of circulation within the jet stream, the activity will be quite a distance to the east and southeast of the surface low-pressure center, which is probably east-northeast of the low aloft; it is from such situations that the classic and devastating tornado patterns develop in the springtime. So when the briefer says "thunderstorm," ask him to look at the 500-millibar chart and tell you if there's a low center shown there. If there is a low aloft over central Texas, for example, and you are flying across Arkansas and Tennessee, beware. Any forecast of severe thunderstorm activity is likely to be accurate.

## EMPHASIS

In leaving thunderstorms, I have things to emphasize.

First and foremost is avoidance. Understanding the storm is an integral part of this, because if you understand how one is built and what it does to an airplane in flight, you'll be properly motivated for the avoidance role.

Second covers some rules of thumb. I'll stick with the lower-is-almost-always-better rule until it is disproven. Stay some knots under turbulent-air-penetration speed when in the vicinity of a storm, because you'll almost always see an airspeed increase as the area of turbulence is reached. If you make that bad mistake and wind up in what appears to be a cell, the best way out is usually straight ahead. Remember that the upset, the loss of control, is the primary problem. Lateral troubles generally precede longitudinal problems, so keep the wings level. In many retractable-gear airplanes it is advisable to extend the landing gear for flight in the area of turbulence, simply because it creates drag, and the more drag, the less rapid the speed buildup in case of an upset.

In checking weather before a flight, ascertain that you'll have a storm-free path for the beginning of the flight. The best information comes from weather radar equipment. Next best is from teletype radar charts or the radar summary chart. The charted information is always dated; be sure you get the time the information was current and project it to the present time. Check the tops: The higher the tops, the meaner the storm. Relate the radar information to other available weather information. Study the weather map. A low-pressure trough can often spawn a collection of storms. Warm front or cold front? Storms associated with the latter are both more organized and more severe. Is there a low-pressure to the west on the 500-millibar chart? Will there be a jet-stream effect on the activity? What's the weather outside the thunderstorms? Is the activity scattered, broken, or solid? The last two can be bad news. Is the level of activity increasing? If it is increasing, surely things will get worse before they get better. Are severe storms forecast?

Once the decision to start out is made, the situation has to be continually evaluated. In a slow airplane it's okay to take it 20 miles at a time. Is it okay in the 20 miles just ahead? If not, in which direction would it be okay for the next 20 miles? Quiz the controller. If you can see, remember that human vision is the best storm-avoidance system available. Continually check weather along the way. If there is doubt, switch to an alternate plan of action.

A final thought is that thunderstorms are dynamic. They build and fade, and they move across the countryside. Often the most difficult thunderstorm situation imaginable will change its character in an hour, or in two hours. The pilot who loses the argument with a storm is the pilot who persists. Often the *only* solution is to land and relax while the storm does its thing in the airspace and moves off to harass the route of some other pilot.

# 9

# ICE

As has been often noted, the classic picture of an IFR pilot
falling victim to ice begins with the aircraft struggling to
maintain the minimum en route altitude as craggy peaks
reach hungrily for the tender aluminum belly. The great fes-
toons of ice continue to grow, and finally the airplane is
overwhelmed. It's an impressive vision, but ice accidents
don't always, or even often, happen that way. The IFR pilot
who lets ice get the upper hand usually loses the battle later
rather than sooner, while doing something that is relatively
simple and that could easily have been done successfully. For
example, a surprising number of accidents occur after the
pilot has extracted the airplane from the icing condition and
as he is maneuvering the iced aircraft for landing. At that
point the stall-speed-increasing properties of ice hold sway,
and the event ends as a stall-spin accident.

Ice, like the thunderstorm, is well respected by general aviation pilots, and that is probably why it is not a direct and overwhelming factor in a lot of accidents. Too, ice seems somewhat easier to manage than thunderstorms, because detection of it is easy; ice allows us to trespass, sniff around at the situation, and leave. It also comes in a better variety than thunderstorms. Ice is light, moderate, and heavy. Thunderstorms are bad, worse, and impossible. The thing that we learn quickly about ice is that it need only be treated promptly to be managed.

Given a proper respect for ice, a pilot will always react to any ice accumulation with an action that will move the airplane out of the icing condition. This might mean climbing, descending, or retreating. The key is in accepting the formation of ice as a mandate. The pilot who watches it accumulate and hopes that it will go away is the pilot who is likely to gather enough of the frosty stuff to have a real problem.

Ice is illogical, anyway. If the temperature is below freezing, you'd just naturally think all the liquid in the atmosphere would be frozen. But a phenomenon known as supercooling defeats the logic. Even though the temperature is below freezing, the supercooled water droplet remains liquid until it is disturbed. Then it freezes. If we charge through a collection of supercooled water droplets in our airplane, disturbing their reverie with complete abandon, they will take umbrage at our splattering them and freeze in retaliation. The temperature and the cloud formations have a lot to do with the nature of the resulting ice.

## RIME ICE

In stratus clouds, the droplets are small, and the small droplets will freeze and become ice crystals if the temperature drops to a value well below freezing. Thus we come up with one of the rules of thumb for evacuating an icing situation: climb to colder air. The ice that does form in stratus clouds, usually rough in texture, is called rime ice, because it is a collection of very small water droplets that freeze quickly, that is to say without splattering all over the place.

Stratus clouds don't have great vertical depth, so any icing condition that is found in pure stratus clouds can often be handled by climbing to on top or to a colder level where the supercooled water droplets have become ice crystals. The important thing is to do it. Even a very light ice accumulation is something to flee.

Next, you might logically ask how one determines the nature of the cloud in which he or she is flying. There's no "welcome to stratus" sign on the clouds, but it is relatively easy to tell one cloud from another. There is no vertical development in a stratus; thus, the air should be smooth. Also, the water droplets are small, and this fact is quite easy to note by just glancing at the moisture flowing on the windshield. The rivulets are fine. If ice does accumulate, it takes the form of a rough and rather milky-appearing substance right at the leading edge of the wing.

## STRATOCUMULUS

When considering cloud types and ice, it is all downhill after the stratus. The stratocumulus cloud likely has many of the properties of stratus in the lower portion of the cloud, but as you climb higher the cumulus part holds sway. There is vertical development and greater depth, and larger super-cooled water droplets are encountered.

Ice formation in anything related to a cumulus cloud is serious, because the accumulation rate is likely to be more rapid than that in stratus. The lifting that makes the cloud a cumulus results in bigger and juicier droplets that maintain their supercooled status to outside air temperatures that would have long since frozen the supercooled droplets in stratus clouds.

If we start off in the bottom of a stratocumulus with the idea of climbing to get above icing, the accumulation might well start off in the classic stratus rime icing pattern. Higher in the cloud, though, the droplets will be bigger. The accumulation might then start taking the form of clear ice, which comes when we burst really big supercooled water droplets. Instead of freezing in a tiny and almost ball-like shape at the leading edge, the droplet splatters, and the clear ice forms both right at the leading edge and back a ways, as the water flows some before freezing. The closer the temperature is to freezing, the farther the water will run back before turning to ice.

Two things introduce us to the problems of the stratocumulus. One, the air is generally a bit bumpy. Two, the droplets are bigger. Just look out at a wing for proof of that. The kicker in a stratocumulus layer comes as we get close to the top of it. There we'll likely find maximum supercooled water droplet size and quantity, the maximum rate of ice accumulation, and the most turbulence. If the plan for ice was to climb on top, and if there is a suspicion that the clouds are stratocumulus, watch out. The tops might well be about 10,000 feet even in low country, and if the airplane isn't turbocharged you are going to reach the level of maximum ice accumulation with a partially frozen bird and an engine that is capable of producing only about 75 percent power or less.

## BEHIND THE FRONT

Some pretty classic ice accumulations can be found in stratocumulus that form behind slower-moving cold fronts, and that cover a wide area. We talked about the pitfalls of climbing in such a situation; if you happen to be flying up over such a cloud deck, beware an altitude assignment on descent that will put you in the upper part of the cloud formation for an extended period of time. The clouds look innocent enough from above. The tops of a stratocumulus layer look almost flat. But there can be plenty of ice in there. It is likely to be cold on the ground, too, so you might have to land with any ice that accumulates during the descent. The message is to work toward an unrestricted descent, at least through the top few thousand feet of the stratocumulus.

## CUMULUS

From stratocumulus into cumulus is like jumping from the frying pan into the fire. While pure cumulus clouds are not likely to be continuous—you'll be in and out of them—the ice buildup can be rapid when you are in cloud, as the airplane encounters supercooled water droplets of maximum size. Some notable collections of wintertime cumulus are found over mountainous terrain—caused by the lifting effect of circulation over the ridges—and this results in a confrontation with ice at the least desirable place. If you are flying on top to avoid icing, and are approaching mountains, you can bet your sweet tooth that the tops will likely become higher over the mountains, too.

It's quite easy to tell when flying in a cumulus. The bumps tell the tale plainly, and you can look at the wing and note the rapid clear ice buildup as the airplane passes through the cumulus.

When operating in an area of cumulus development, it is tempting to try to rationalize some hope into the situation as the airplane passes from areas of icing through areas with no icing and then back into the areas of icing. The fact that the buildup is not continuous should not be used as an excuse to carry on, though. The sporadic buildup in areas of cumulus can easily exceed the buildup found during continuous flight in stratus clouds.

Unless the airplane has exceptional altitude capability, the odds on climbing out of a cumulus icing situation are quite against you. An unturbocharged airplane would have little chance in such a situation. The way out is almost always to warmer air below, to a position beneath the clouds, or back to the air from which you came.

One of the principles of meteorology does work in our favor. Cumulus clouds are found when the air is unstable; when the air is unstable it cools more rapidly with altitude. In reverse that means it warms more rapidly as you descend, so the chances of finding warm air beneath are better than in a stratus or stratocumulus icing encounter, for example. No guarantee, though, especially in rough terrain, where there might well be no warm air above ridge level when cumulus clouds are producing large amounts of ice.

Clouds tend to mix it up and become difficult to identify in many situations, but the rule of thumb about icing in cloud being worse when there is turbulence is a good one. For example, when flying in the quadrant to the northeast of a low-pressure system, especially a developing low-pressure system, there is generally both instability and moisture. You might not see a textbook picture of a cumulus cloud when flying in such a situation, but you can sure experience the ice and feel the bumps.

## ——— . . . AND IN PRECIPITATION ———

The weather people always forecast icing in clouds *and* in precipitation when conditions are ideal, and there is one particular form of precipitation to beware. Freezing rain is a creator of often-mentioned fantastic ice accumulations in a very short period of time. This occurs when rain falls from warm air above into cold air below and splatters and freezes when it contacts anything in the below-freezing air.

The accepted procedure when encountering freezing rain is to climb into the warm air above. This works, but it must be done very quickly, and a better answer is often a move back out of the area of freezing rain. If you climb up to warmer air above and then have to make an approach to an airport where freezing rain is in progress, the maneuvering for landing will be in one of the most severe ice-producers of all. As you descend and maneuver so much ice might accumulate that the airplane won't be able to climb back to warm air above if the approach is missed. That puts quite a premium on doing it correctly the first time, and falls in the high-risk category.

There's ice to be found in snow, but this is usually not too much of a problem. Dry snow just passes on by without sticking. Dry snow can cause precipitation static, and this can play hob with communications and navigating. Good antennas and static-discharge wicks should eliminate this, though.

Wet snow can stick, and when you're flying in a wet-snow condition the temperatures aloft become critical. Right at freezing is bad. Too, wet snow comes in big flakes, big flakes are generally associated with cumulus-type clouds, and cumulus clouds are associated with clear ice. In such a situation, the pretty snowflakes might not be the actual creator of the ice, but they might well be the signal that the clouds will be thick and icy.

## *THE WORST*

If I had to pick the worst possible icing situation, I'd go back to the beginning and pick that airway in mountainous terrain, with the airplane accumulating ice while flying at the minimum en route altitude, which also happens to be at or about the airplane's service ceiling. That situation is the worst because it is without options. We *must* have options when ice starts forming on an airplane. When you have none, you are good as whipped.

From that optionless stereotype, move on to the more likely ice encounter and see how the cards might be played.

Fly into an icing condition in cloud during climb, soon after takeoff. This one is pretty simple. If it doesn't go away in a couple of thousand feet of climb, return and land. A return to mother earth can be the most attractive option of all.

En route, fly into an icing condition at a moderate cruising level—say, 6,000 feet. The options here are wide, even without turbocharging: higher, lower, retreat, or continue without change. The last is the poorest choice. If the situation went from one with no ice to one with ice, the chances of its getting worse before it gets better are quite good. Some basic change in the situation created the ice, and it's just not likely to go away quickly.

Higher might be the best option if the air remains smooth. If a definite tops report could be obtained, and if that level is easily within the capability of the airplane, climbing would be by far the best deal.

A lower altitude would depend on the surface temperatures and the minimum en route altitude. If the country is flat, and MEAs down to a substantially lower altitude are available, that might be a pretty good deal if the temperature at 6,000 is just below freezing. Another consideration on lower altitudes is the availability of places to land. If there are adequate airports with approaches (and minimums) beneath, the lower altitude is always a good deal.

## KNOW THY SITUATION

Some basic knowledge of the weather situation can be applied to this situation, too. For example, if you are flying toward colder air, level at 6,000, and ice starts forming, any descent to a lower altitude might provide only temporary relief. The temperature at 4,000 might soon drop below freezing. And the temperature at 2,000 might do the same as some more miles slip by.

Watching the surface temperatures is quite important if the decision is to descend, but don't kid yourself with surface temperatures that are close to freezing.

The temperature can drop as rapidly as 5 degrees per thousand feet, though the temperature in cloud drops less rapidly than that. At any rate, a surface temperature of 34 or 35 degrees F is nothing to consider as much of an option. Even with a surface temperature as high as 40, the minimum IFR altitudes might all be icy, even over flat terrain. You really need to correlate surface temperatures with what is being experienced at the cruising level to get some picture of available relief at lower altitudes. If you are flying at 6,000 over flat terrain and it's freezing, and if the surface temperature nearby is 39, you'll probably be able to do some good by descending. But remember that it might not last if you are flying toward colder instead of warmer air.

For further illustration, assume that our response to an ice accumulation was to climb instead of descend. The climb might well be a long one, but let's say that we reached either an on-top situation or a temperature too cold (at least –10 C, more likely –15 C and even lower in cumulus clouds) for icing at 12,000 feet. Now things are better—no more ice is forming—but still not ideal. The airplane has some accumulation and little more climb is available (unless the airplane is turbocharged). There will thus be one less option if ice starts to build again.

## E-FLAT

Here it might well be noted that ice accumulations can make an airplane emit a strange collection of noises. Some practically shriek when iced. Fixed-gear Cessnas certainly do this, and I've always thought it must be some result of ice formation on the nose gear. At least that's where the sound seems to be coming from. Also note that the ice affects airspeed, and the range of the airplane will be cut. Ice on a maximum-range flight is good reason to review the fuel reserves.

Those two factors acknowledged, move now into further icing at 12,000 feet. More is accumulating; what shall we do about it?

With the options reduced, a pilot in this situation has to face a lot of basic facts. If the airplane is not capable of climb, then much more ice accumulation will probably make it incapable of staying at 12,000 feet. What is below? If there is warm air below, fine. If there is a between-layers situation below, fine. Go for either, but go with an out. Maybe that lower altitude isn't as warm or cloud-free now as indicated on a pilot report a few hours old, or maybe the forecast is wrong. Don't count on anything unless there is a brand-new pilot report to back it up.

When leaving 12,000 feet for alleged warm or cloud-free air below, I'd try to add the option of an airport with an instrument approach and comfortably above-minimum weather conditions available nearby to use as a haven in case the prediction of good things turns out to be a fiction.

What if there is no airport below, and no hopeful situation below? Well, you can spend a millisecond cursing the fates, or your own judgment in bringing the airplane this far, and then go on to the best possible option. I would fly the airplane at the highest altitude it would remain until within striking distance of an airport that I felt reasonably confident of hitting on the first instrument approach. Then, and only then, would I start intentionally descending. Altitude in this situation is the *only* thing in your favor. It is stored energy and can be translated into distance. Don't give it up prematurely.

An airport for such an arrival should be chosen with care. Ideally, it would have an ILS and reported weather well above ILS minimums. I would arrange for a straight-in approach, no circling, and would try for about a ten-mile turn on final. I think that I'd leave the gear up on a retractable until I had the airport made, I'd use flaps according to the pilot's operating-handbook recommendation for icing conditions (usually no flaps, sometime partial flaps), and I'd keep the airspeed 20 to 25 knots above the normal approach speed until runway is beneath the airplane.

That is, of course, assuming quite a collection of ice. And you'd have quite a collection after flying some miles in a gradual descent from 12,000 feet in icing conditions all the way down.

## HIGH COUNTRY

There are some areas where 12,000 feet isn't a lot of altitude. One of the higher minimum en route altitudes in the country is between Pueblo and Gunnison, in Colorado. The chart says that 16,000 feet is the minimum; fly it and you'd say that 20,000 feet or even more would be a lot more comfortable. There is a lot of vertical real estate, and perhaps the best way to avoid ice problems along such airways is to refrain from flying IFR along them unless in a turbocharged airplane, preferably with de-icing equipment.

Even turbocharging is no cure-all, because the turbocharged airplane is, in the high country, in much the same shape as the unturbocharged airplane out in the flat country. It'll operate perhaps 10,000 feet above the minimum en route altitude, so a turbocharged pilot approaches ice in the mountains with the same vertical limitations that the nonturbocharged pilot finds between Chicago and St.

Louis. One big difference: Between Chicago and St. Louis you might settle into a farmer's field with an iced airplane. There aren't many such fields in the Rockies.

## ——————————— *HARDWARE* ———————————

De-icing equipment is good stuff to have if you fly a lot of wintertime IFR, but there is certainly no magic to it. Just as radar won't allow penetration of thunderstorms, de-icing equipment won't solve every ice problem. In heavy icing situations, enough can accumulate on surfaces that aren't de-iced to cause a considerable problem. I've even gotten enough ice accumulation on heated props to cause quite a bit of vibration. (The ice formed right next to the prop hubs in extremely cold temperatures close to the top of a stratocumulus deck.) What de-icing does is expand the time available to choose one of the options available for fleeing the icing situation. A pilot would be purely foolish to remain at an icing altitude and depend on the de-icing to keep things completely cleaned off. It just doesn't work that way.

There's long been an argument about which is best to de-ice—props or flying surfaces. Naturally, it is better to de-ice both, but for many of us there's no way to de-ice the surfaces. Boots are presently the only way, and boots are not approved for most single-engine airplanes. Perhaps the unavailability of boots plus the availability of a simple deterrent to prop ice is what has led so many of us to say that keeping the prop clear is the most important item. Self-deception, perhaps, but keeping the prop clear is better than nothing, and it does avoid vibration caused by a little more ice on one blade than the other.

There are some ice-deterrent preparations available in spray cans, and these have served me well on propellers. I keep my prop filed silky-smooth, and I squirt some of that stuff on the propeller blades before each flight that has the slightest chance of passing through any icing condition. And I have never had any problem with propeller ice. (The knocking sound you hear is four knuckles on the piece of wood that I always keep handy.) Many pilots also spray the antennas, and I've seen some spray the leading edges. I've done the latter but have noted no advantage to it. The stuff might keep ice off, but only for a minute or two. The reason that it works well on the prop is related to centrifugal force and propeller-blade flexing and vibration coupled with a slight lingering slickness of the preparation.

## CAN YOU SEE?

Unless you fly a de-iced airplane the windshield can be quite a problem in icing conditions, because the average defroster isn't up to keeping even a small portion of the windshield clear. I have found some small degree of success in putting a leather Jeppesen book on top of the panel, almost over the defroster opening and as far forward as possible, to hold heat in an area at the base of the windshield. If the icing isn't too bad, this might keep clear a peephole to use in landing. If the defroster puts out extremely hot air, though, beware of getting the inside of the windshield hot enough to soften the plastic. Too, if icing is severe enough you could probably get the inside of the windshield quite warm without keeping the outside ice-free.

If the windshield does become opaque with ice, you'll learn that the little storm window on the left side is for something other than summertime ventilation. You can open it and at least see something out to the side and forward. Some are better than others. Fortunately, it isn't too difficult to land a light airplane when you can't see straight ahead. Preferably an icy landing would be on a nice big runway with an ILS to help provide guidance to the touchdown zone.

I do not think that there are many hard and fast rules to use in making the preflight start/don't start decision as it relates to icing conditions. Perhaps the primary thing is in examining options. Surface temperatures are important. So are temperatures aloft, though these will more often than not be forecasts. Pilot reports are good items of information if they are reasonably current. As with thunderstorms, the important thing is to insure that it is okay to start out and that there are options from the beginning. In a freezing-rain situation there might be no such options. With cloud bases of 1,000 feet, surface temperatures 5 degrees above freezing, and icing forecast in the clouds, you might have a look with the option of returning.

The weather map is important to preflight icing deliberations, too. Is there something out there that will create lifting and cumulus clouds? Warm front, cold front, or terrain effects? How about a low-pressure to the west or south? Some of the classic icing situations in the central and eastern U.S.

occur as a low-pressure area moves by to the south in the wintertime, bringing moist air up to mix with the cold air from the north country. Brrr.

## ATC

Another ice item is related to air traffic control. As soon as ice is encountered, tell the controller and ask for pilot reports. If your decision is then to climb or descend, or to retreat, preface the request with the reason: "Center, 340RC, due to icing I'm requesting 8,000." If the destination is changed because of ice, say why. And if it becomes like a situation related, where it was suggested that the airplane be kept as high as possible as long as possible, make the request and explain the reason. Air traffic control procedures might normally call for a descent to 2,000 feet 10 miles from the airport, for example, when you'd be a lot better off at 3,000 feet and descending when 10 miles away. Don't stay up so high that you'll overshoot, but plan on a final descent rate of not less than 300 feet per mile traveled, or 600 feet per minute at 120 knots groundspeed.

Ice is neither mysterious nor complicated, and when a pilot actively wonders about the most dangerous thing about ice there is but one answer. Procrastination is the prime hazard. Eliminate the procrastination and you might not always get where you wanted to be, but you'll have a lot better chance of always landing at an airport when ice becomes a problem.

Too, good knowledge of the weather picture can avoid surprises. Just knowing whether it is warmer, colder, or about the same in the area toward which you are flying can tell you a lot about what to expect. Watch the temperature trend on your thermometer, too. If the airplane is moving along in cloud and the temperature has slowly dropped until it is at the freezing point, don't be surprised at the ice that forms. And don't forget that turning around and returning to the warm air behind will make it go away.

# 10

# DOING IT
# IN THE DARK

First, acknowledge two things about flying instruments at night. One, it is something that most of us do not do with regularity. Two, the accident rate in IFR flying is very much higher at night than in the daytime.

Next, add one more fact: The involvement of actual mechanical engine-failure in single-engine airplanes during night IFR flying is so insignificant that it will not be considered here.

In some first-night-IFR experiences, almost every pilot must occasionally have that old "What am I doing here?" feeling. I still do at times, and I suppose that it comes from a realization that night IFR requires stronger discipline than any other form of flying. The work that we must do is more difficult. The airplane doesn't matter so much. I have the feeling in twins just as strongly as in singles. There's just something different about night IFR.

Start with the end of a flight and compare a daytime instrument approach with a nighttime instrument approach. By
day, if there is an 800-foot ceiling, the approach might well
be a piece of cake. We'll be in visual flight conditions at almost a normal traffic pattern altitude, and the final landing
approach, whether it be straight-in or circling, becomes a
normal VFR operation. Even somewhat lower approaches in
the daytime offer visual flying toward the last, with relatively
reliable clues. It's not until we get down to the really low
ones—ceiling 300 feet or below and visibility one mile or
below—that the daytime approach starts offering visual bear
traps.

## BY NIGHT

It's entirely different at night. There are *no* reliable visual
clues on a night instrument approach. This holds true even if
the ceiling is at or above normal pattern altitude. This was
once illustrated in a study by a manufacturer of air-carrier
aircraft. They had experienced pilots "fly" a night-visual-
approach simulator (no instruments) that offered the opportunity to fly to and land on a runway, heads up all the way,
strictly visual. It was found to be very difficult to avoid landing short, in the rough, especially with certain combinations
of ground lighting.

The situations in which it was most difficult to judge
height accurately were: a long straight-in approach to an airport located on the near side of the city; a runway length-
width relationship that is unfamiliar to the pilot; an airport

situated at a lower elevation and on a different slope from the surrounding terrain; approaches from a navigational facility located some distance from the airport; substandard runway lighting, and other landing aids not available; a sprawling city with an irregular matrix of lights spread over various hillsides in back of the airport; industrial smoke or other obscurations that make lights look dimmer or farther away.

As we make an approach to a strange airport at night, there's no way to know that one or more of these visually misleading factors won't be present, so it is best to assume that a trap will be there. Add another one to the list, too. Precipitation often gives an illusion of greater than actual height. Whenever it's raining or snowing, be especially wary of what you *think* you see.

## ILS

If an airport has a full instrument landing system, the glideslope provides vertical guidance to the runway and should be followed until the runway is beneath the wheels. In the daytime, it's important not to change anything when becoming visual on an ILS approach—just maintain the rate of descent and power setting that had been tracking the glideslope and don't reduce power until time to flare. At night this is even more important. If pilots are prone to go below the glideslope when changing to visual flight in the daytime, they are even more prone to do so at night.

Approach lights can be a trap at night, because they become visible sooner than they do in the daytime, and they make an offer that is not really there. The approach lights seem to penetrate cloud and say: "Come on, we'll lead you down the runway." It is true that they will lead you on in a left-right sense, but they offer absolutely no up-and-down guidance. Only the runway itself can offer a valid cue for visual-approach-slope judgment. You have to see the aiming point to aim, and if the approach lights are accepted as offering anything in the vertical sense, they become the aiming point. Drive in your car to the approach zone some day and look at the poles on which they mount those approach lights. Very uncomfortable aiming points. And unless you take approach lights as an item of incidental information, you run a good chance of tangling with those lights and poles. One light system that does help lead you to a proper point in the up-down sense is the visual approach slope indicator. The VASI is invaluable on a night approach, and it is too bad more small airports don't have this.

## NON-PRECISION

Moving from the full ILS back to a non-precision approach (one without electronic vertical guidance), to a runway without a VASI, we find ourself in the aeronautical dark ages. No pun is intended. In recalling the things that can create erroneous visual impressions, we know that there is no way to count on making an accurate approach visually, and we know that there is nothing on the instrument panel that will calculate a proper approach slope for us. The pilot's difficult task begins at the time the airplane leaves the safety of minimum descent altitude, and it becomes one of very careful interpretation and interpolation.

A long final to an airport in the distance can be quite demanding. I recall making one such approach to a runway that was much wider than I was used to. A feeling of uncertainty crept into the proceedings when I was a couple of miles out on final. Or was it 3 miles or 4? According to the altimeter I was only 800 feet above the airport, I thought the runway was pretty far away, and yet the urge was to descend. The visual suggestion was one of being too high. I just could not quite resolve the problem, so I gave up on my straight-in approach. I remained at the MDA listed on the chart for a circling approach, flew up over the airport, made a normal pattern, and landed. By so doing, I was at a known position when I started my descent. When I was over the airport, I knew right where it was. That was a lot better than the previous situation of flying over an inky abyss with runway lights an unknown distance away.

## ———— JUDGING THE SLOPE ————

There is a valid technique for judging the proper approach slope and I will repeat it here.

A 15:1 approach slope will clear all obstructions on a VFR or IFR runway. So decide that 15:1 will be the minimum acceptable slope for your approach. That's 15 feet forward for each foot down. At 90 knots groundspeed, we are moving forward at a rate of about 9,060 feet per minute. Divide by 15 and you'll find that 600 feet per minute would be a minimum allowable rate of descent when the groundspeed is 90 knots.

The point on the ground toward which the airplane is tracking remains in a constant spot in the windshield. This is somewhat easier to see at night than in the daytime because of the runway lights against a generally dark background. If you are making good a descent toward a point a light or two past the approach end of the runway, if the groundspeed is 90 knots, and if the rate of descent is 600 or more feet per minute, then you know that your approach slope is 15:1 or steeper. If the airplane were tracking toward the runway at 90 knots with a rate of descent of only 300 feet per minute, the slope would be dangerously shallow.

So there is information on the panel to use. It isn't as good as a glideslope, to be sure, but it is sure better than nothing. The altimeter, airspeed, and vertical speed all give solid messages; combined with visual observations, they can help keep the airplane out of the trees and hills. And while these instruments are important when maneuvering visually in the daytime, their messages are absolutely essential at night, when it is at least somewhat IFR all the way to touchdown every time. Think in terms of steep approaches with minimum acceptable rates of descent to track to the end of the runway. If the slope is too shallow, fly level or even climb until closer and then resume a descent.

## ———————— CIRCLE TO LAND ————————

In the mention of a flight up over the airport for a circle instead of sticking with a straight-in one dark and lonely night, I contradicted a long-held feeling that a circling approach to minimums at night in turbulence and precipitation is the most difficult maneuver that can be attempted in an airplane. What I was doing in the switch to a circle that night was moving away from something without points of reference—the long straight-in—to something with points of reference. On that flight it might have felt more comfortable, but that didn't make it easy.

The circle at night is difficult because the discipline on altitude control is absolute, visual factors might try to make you fly lower than you should, and there is no way to see where you are going.

The approach plate does tell us how to stay out of trouble while flying a circling approach. Say the MDA is 458 feet above the ground and the visibility minimum is one mile. Follow the published final approach course while flying at the MDA until within one mile of the airport. Then start the circle, remaining within one mile of the airport. I would not leave the MDA until turning onto final for at 90 knots the required rate of descent to lose 458 feet in a mile is a bit less than 700 feet per minute—just about right. That rate of descent is quite easy to handle in a light airplane, it gives a satisfyingly steep approach, and it leaves me at the safe haven of MDA until I can draw a bead on the aiming point.

There are a lot of reasons why it is best not to leave the circling approach MDA until turning final.

The rules say not to leave MDA unless in a position from

which a normal landing can be made, and unless the runway, approach lights, or other markings identifiable with the end of the runway are clearly visible to the pilot. What the rule does not say is that you shouldn't leave MDA unless you expect these things to remain true. At night, there's no way to see cruddy scud between you and the runway until you turn final and take aim on the runway. There is no way to assess the flight path ahead and make a determination that it will indeed be possible to fly visually onto the runway unless you are looking down your approach slope to that runway. Only after turning final are you sighting through the air between you and the place you want to be. So only after turning final should you leave a known safe altitude.

In high-performance airplanes, the descent rates on final required after flying a circling approach in this manner might be not too desirable. In such a case, it might be best to stick with a higher-than-published MDA and make a wider circle, still to leave MDA only after turning final. If that isn't possible, surely there's a nearby airport with an ILS. The accident history in low-visibility circling approach at night is so very bad that it is surely unwise to try very hard to complete such an approach in doubtful conditions. Be aware of this, and consider the event as a very-high-risk maneuver.

───────── *GO UP, YOUNG PERSON* ─────────

The following is true day or night, but it seems to be a more difficult point of discipline at night so it is emphasized here. If, after leaving MDA or DH, the view of the runway is lost, or even becomes the least fuzzy, take this as a mandate to go *up*. It is never a signal to descend, and pilots who descend and try to maintain visual contact with the runway after flying into scud are those most likely to fly into the ground on day or night IFR approaches, especially night IFR approaches.

Discipline is more difficult at night, because where we have only a gray mass scooting beneath in the daytime, lights beckon at night. Lights tend to penetrate cloud, they tend to suggest that you come on down, they tend to suggest that the visibility is better than it really is.

The night approach isn't easy, and its difficulty combines with visual illusions to make it a critical area for risk management. Take no chances. Follow the book to the letter and the altitudes to the foot.

## PREPARATION

Backing up, preparing for the approach, and studying the paperwork is more difficult at night, because cockpit lighting is never ideal. Too, as you add a touch of age, your night vision is the first thing that suffers. You can walk out of the aviation medical examiner's with a fresh physical and still notice some difficulty in accommodating to the visual demands of night IFR.

Oxygen helps night vision, and I find it quite refreshing to spend a while on oxygen before leaving altitude for terminal-area maneuvering at night. A small flashlight is essential for chart reading, too, because, few map lights direct a bright stream of light at the chart.

The things that we do to prepare for an approach in the daytime need special emphasis at night. If you aren't a frequent night IFR flyer, getting behind is both easier and more serious. The arrival needs to be planned carefully, and any sign that the flight isn't following the script is a clear call for a reassessment of the activity.

I'll always remember hearing a pilot come to grips with reality one bumpy night. He missed a VOR approach, and was working his way around for a second approach when he told the controller that he'd just like to change plan and go to another airport, one with an ILS. The pilot recognized that he was not organized, that he was pushing a non-precision approach, and that he'd feel better with more electronic guidance. Wise man.

I also recall a night approach I shot that involved special pressures. I was in a twin, they held me up above 10,000 feet until I was practically at the destination, and the descent was made in an abbreviated holding pattern. "Just do 360s while you descend." By the time I leveled at the glideslope-interception altitude and started inbound on the ILS I was pinching myself and doubling up on the challenge and response (all by myself) to make certain that the required disciplines were intact for the approach to minimums. I would have much preferred a "normal" arrival, but you have to adjust to actual situations, and in this case I found that a lot of extra effort was necessary to make the adjustment.

## ALONG THE WAY

En route, the thunderstorm is worthy of a special night thought because there are a number of thunderstorm problems involving IFR night flights each year. It is hard to pinpoint a reason why night would offer more problems than day in relation to storms. Some feel that being able to see lightning should actually make the night avoidance task easier, but apparently this is true only when avoidance means staying a gross distance away from any lightning discharges. Perhaps the ability to see lightning leads pilots to use it as a visual aid for penetration, and the clues from lightning are far from adequate to use in avoiding cells at close range.

Also, it is often noted that strobe lights and flashing or rotating beacons should be doused when flying in cloud at night. I think that most of us have turned them on just to sample the effect, or have left them on for a moment after flying into cloud. The results can be spectacular. I doubt that they really cause spatial disorientation; rather, the illumination of the cloud's innards is so bright that it probably draws the eye and makes you neglect the instrument scan. Same result.

## TAKEOFF

Moving back to the departure, there are some special considerations for night IFR takeoffs.

When we launch in the daytime, the ceiling can be "seen"

if there is a ceiling. On a day with, say, a 600-foot ceiling, I might fly visually up to the point just before cloud penetration. Then I change over to instruments. That's perhaps not the best way, but when you can see it is hard to resist looking right up to the last minute.

At night, there's no way to clearly see a ceiling, and it is not wise to depend on the reported value to decide when you are going to have to change to instruments. At night, I turn the panel lights up full bright, make the takeoff without landing lights unless there is some compelling reason to use them, and then switch to instruments at liftoff. This way, I'm ready for the clouds at whatever level they choose to envelop the airplane.

The sensations of a night takeoff and initial climb can be more bothersome than a daytime IFR departure, and a dedication to the gauges helps relieve the discomfort. In many situations, regardless of the type of flight plan, the flight instruments give the only valid visual clues on a night departure—especially if launching from a well-lighted airport and flying out over a dark or sparsely lighted area. A definite transition to gauges can make the first 1,000 feet of climb a much more precise exercise in such conditions. If you must, after reaching that altitude look outside if it appears the airplane is in visual meteorological conditions. Look for other traffic and think you are flying visually. But scan back to the gauges, because only they give the straight poop.

───────────── *PREFLIGHT* ─────────────

There are some special preflight actions that can help make night departures more routine. The business of arranging the first part of the flight—presetting everything that can be preset and knowing the path and altitude that will be prescribed for the first part of the flight—is important, because where you can glance at a chart-on-lap by day to refresh your memory, at night you have to arrange for lighting and then *squint* at chart-on-lap. There's a big difference.

Have the flashlight ready on the takeoff roll, too. I learned the value of this one very dark and rainy night when departing from a 3,000-foot-long runway with minimum runway lighting and virtually no ground lighting around the airport. The airplane was a light twin, we were at gross weight, and as the aircraft reached liftoff speed pretty far down the runway the panel lights flickered a couple of times. I mentioned to the lad riding shotgun that he should have the flashlight ready. We were just off, gear coming up, when the panel lights went out entirely. It was a bad moment. The airplane was just beginning to climb, my scan had been arranged to hold a heading, a pitch attitude to insure a positive rate of climb, and to verify a proper airspeed. Then there was nothing but darkness, the drone of a couple of Continentals, and the pitter-patter of rain on the windshield.

Fortunately my flying companion of the evening was fast on the draw with that light. I asked him to shine it on the artificial horizon. Everything was still okay, and I left the engines at full-bore and climbed at the best rate-of-climb speed to a good altitude. I had a flashlight in my pocket, too, and I wasn't particularly concerned about running out of bat-

tery power, but the view of the panel was somewhat restricted. And I felt rather dumb when one of the passengers,
who had watched the whole procedure, wondered aloud if it
would help us to turn on the dome light. He pressed the but-
ton. It did a fine job of lighting the instrument panel, and we
flew on home. We'd have been in a difficult situation at the
moment of lighting failure had we not had the flashlight
ready, though.

## WEATHER

There's a weather-related item to watch at night, too. Generally ceilings and visibilities will drop at night in an area of
inclement weather. And the reporting of weather at night
seems to me very much less accurate than the reporting of
weather in the daytime. It is not unusual to find yourself flying to the final approach fix at night, 1,500 feet above the
ground, in cloud, with a reported ceiling of 5,000 feet or better. The observers are looking at instruments and a lot of
black sky at night. In the daytime they can see clouds, and
are less likely to miss the fact that lower clouds are banked to
the west of the airport, for example.

One final item on night IFR. It was noted in the beginning that mechanical engine failure is statistically insignificant in nighttime IFR accidents, and that we thus wouldn't
discuss any pros and cons of the one-engine-versus-two question. There is an engine-failure question that must be noted,
though. It is not related to things breaking, but to fuel. There

are cases of fuel exhaustion on night IFR flights, and occasionally a pilot will position a fuel selector incorrectly and wind up with a dandy case of fuel starvation while trying to juggle all the balls of an instrument arrival. Either event can be quite unhandy.

On fuel quantity, it's wise to be extra super-conservative at night. If you land with an hour's fuel on board by day, land with two hours' fuel on board at night. Weather conditions are more likely to go below landing minimums at night than in the daytime, and if the weather area is of any size, the trip to an alternate could be a long one.

On fuel selection, use a flashlight to illuminate the fuel selector(s) if lighting is not provided. As well as we often think we know an airplane, it is still possible to turn something in the wrong direction. But if you shine a light on it and select "Right," for example, then you'll know that it is indeed on "Right."

Nothing wrong with night IFR, nothing at all. It's all a matter of recognizing the extra demands—there are a lot of them—and catering to those demands. And when compared with night VFR in marginal weather, night IFR is one of the best deals going.

## INTERLUDE—A VIGNETTE:
## EVERYBODY'S GOTTA BE SOMEWHERE

You have probably heard the old story about what the fellow said when he was discovered hiding in the closet. Everyone *does* have to be somewhere, and in instrument flying our location in space is the thing by which we live. A pilot who is always aware of position and altitude, as well as aircraft attitude and speed, is a pilot doing his IFR work properly.

It's often amusing to respond to a passenger's question about location with "Beats me" as we move along in cloud. It had better be a joke, though. When flying IFR, you must *know* exactly where you are and what you are doing at all times. Instead of "Beats me" a better answer might be: "We are on Victor 54, 52 miles northeast of Little Rock, flying at 7,000 feet on an airway that has a minimum en route altitude of 2,500 feet. Our groundspeed is 150 knots, and we'll be over Memphis at 45 past the hour."

Altitude is a critical part of our position in space. When pilots crash on an approach or while maneuvering in a terminal area, it is usually because they are flying too low. There are bad altitudes and there are worse altitudes in thunderstorms. Altitude is an important consideration when dealing with ice. Selection of the best altitude for winds might make the difference between getting there with good reserves and sweating fuel to the last drop. Flying at too high an altitude without oxygen or pressurization can fuzz the mind for a critical approach. Flying at the wrong altitude can create a collision hazard.

First, is the altitude a safe one? Second, is the altitude the best possible safe altitude for efficiency and comfort?

When we fly VFR, we draw the line of the chart. Avoiding terrain and obstructions becomes a matter of noting elevations on the chart and visually verifying that the selected altitude is a safe one. When we fly IFR, the lines are drawn on the chart for us and there is always a minimum allowable altitude for each line. These minimums are not to be taken lightly. There is good air at and above the numbers; below are trees, rocks, and TV towers.

Acknowledge, too, that there is often temptation on an approach to fly at a lower altitude than a safe or legal one. The devil on your shoulder might say: "There is always a ceiling, so it is okay to descend below MDA to an altitude beneath the ceiling." The devil is correct. There is always a ceiling. But it might be halfway up a tree, and just before you fly into the top half of the tree the devil that made you do it will hop off and go find another sucker's shoulder on which to ride.

Altitude and position are the sum of what we do. Knowing both is imperative. "This is my position and altitude, and this is a safe altitude for this position." Ask the question continuously.

# 11

# IFR EMERGENCIES AND GLITCHES

My dictionary defines an emergency as a sudden, urgent, usually unforseen occurrence or occasion requiring immediate action. If they would just take "unforeseen" out, it would be an excellent word to use in aviation in general and IFR flying in particular. As it is, the word "emergency" doesn't quite fit, because "unforeseen" has no place in flying. A person must learn to expect anything and everything when operating an airplane, and must have a plan to use in handling any problem. "OhmyGodwhatdoIdonow?" just does not work in airplanes.

The part of the definition about immediate action *is* applicable, but consider that "immediate" doesn't mean that things should be done before you have taken the time to verify that the chosen action is indeed a correct one.

Start with a simple thing, an event that doesn't threaten

the actual safety of the flight but does demand attention. Right after you punch into the overcast, the baggage door of the airplane comes open and starts distributing the contents, dirty underwear and all, over the terrain below. In such a case it might be tempting to immediately duck back down, get VFR, and return and land. That's no good, though. Once the airplane is in cloud, the *only* thing to do is to follow IFR procedures. In this case, the solution would be to request clearance for an approach back to the airport of departure. Do it routinely. Do not cut any corners. A lot of airplanes have been destroyed after a door inadvertently came open, but in general aviation it is usually precipitous action on the part of the pilot and not the actual opening of the door that causes the accident.

There are numerous other things to come open or come loose on an airplane, and the event is almost always best handled by a routine IFR return and landing. The inconvenience of this is why the IFR preflight should be extra thorough. Landing to close a door when VFR is no big deal. It becomes quite a production on an IFR flight.

## CAN'T HEAR

Moving on, there's a relatively simple thing right after takeoff that often plagues pilots. Somehow, if there is to be communications difficulty, it seems to come more often as we depart. Perhaps the number two radio is not used until we try to talk with the departure controller and the radio isn't up to snuff or the squelch is incorrectly set. Or perhaps the audio switches are positioned incorrectly, or the frequency in the window is simply not the correct one. Whatever, the result is an unanswered call to departure control.

Loss of radio contact shouldn't be considered an emergency, or even an event of air-shaking importance, because there are procedures to study and follow. A pilot should always have a plan in mind for continuation in the event of loss of two-way radio communications. It is a cinch you can't park the airplane on a cloud, and generally the controller is going to expect you to behave as flight-planned. Loss of communication is far less critical in a radar environment, too, because they are watching and will protect a lot of airspace for an IFR airplane with which they can't communicate.

## NAV LOSS

Considering other avionic failures, a rule requires a report on the loss of navigational gear. Don't fail to make such a report, and to ask for any special favors that might seem in order because of the failure. For example, if after losing one VOR you are told to hold at an intersection identified only by two VOR radials, ask for relief. Holding at such a point with one VOR is certainly possible, but the attempt is better avoided. There was once a tragic collision between airliners that was related to an intersection and an airplane with only one VOR receiver, and that lesson shouldn't be lost on us.

Even though the rules don't specifically require it, any other failure—such as of vacuum pump or alternator—should be reported to the controller. The people on the ground should know about *anything* that might have an affect on the pilot's performance in the system.

## LOW FUEL

If you get your hand in the cookie jar on fuel, and it looks like it might be close, air traffic control should be told that the flight is arriving with minimum fuel. That'll help smooth the way. But watch out for cutting corners on any such approach. An extra mile or two on final to get it all lined up consumes a lot less fuel than a missed approach, and a missed approach is a likely followup to an overexpedited arrival.

While making any approach in a retractable, consider what you might do if the landing gear extension system should malfunction. The landing gear shouldn't normally be extended until at the final approach fix inbound. Dragging an extended landing gear farther than necessary is a pure waste of fuel, and wasting fuel is bad regardless of how much is in the tanks. Waiting until this late in the approach does mean that the gear probably can't be extended with the emergency system while you are continuing the approach. If the pilot already has one problem, such as low fuel, a malfunction of the gear-extension system means that the dominoes have begun tumbling. Low on fuel, I for one might well follow the path of least resistance and just continue and land wheels up. That would beat running the risk of encountering still another problem, such as fuel exhaustion, while flying around trying to get the wheels down.

## WATCH FOR FLAGS

The approach is a good time to give some more serious thought to navigational-system failures. The instruments have flags on them to indicate failures, but the flags are often hard to see. Too, we tend to look past flags. Many a pilot has thought that he or she was doing a perfect job of flying a needle when in reality the needle was remaining centered because it was dead. I always set both nav receivers to the ILS frequency when on that type of approach; if ever the needles did not agree, I'd pull up and investigate. Most airplanes do not have dual glideslope receivers, so there we must watch for the flag and monitor altitude for logic. On a VOR approach, I set both nav receivers as appropriate for the approach whenever possible, for a consensus.

## AIR AND LIGHTS

Either of two system failures—vacuum or electrical—can pose serious problems, especially in single-engine airplanes, but either is manageable with a modicum of attentiveness and planning. We've noted in Chapter 2, on partial-panel flying, what happens when the vacuum system fails. If the alternator fails, the critical thing is to catch the failure *when it occurs*. A low-voltage warning light is excellent for this, because it will come on very shortly after an alternator fails. Planning to maximize the energy stored in the battery can then start at the very beginning. If you don't catch a failed alternator for an hour, chances are you'll catch it because things start to fade. At that point, you are left with few options.

How long a battery lasts depends on a lot of things, including temperature, the condition of the battery, and current drain. There's nothing we can do about temperature. The only way we can influence battery conditions is to service it regularly and buy a new one at specified intervals—every year, or two years at most—instead of when the old one just falls over dead. Current drain is our controllable ace in the hole, and if it is minimized from the moment of alternator failure, there should be enough in the battery to last through an approach and landing.

It is very safe to say that a good battery should handle one nav/com radio for at least an hour after an alternator failure. It should actually last much longer than that, but there is seldom a time when there's not a place to land within an hour. An alternator failure is a mandate to land as soon as possible, so there's nothing wrong with a one-hour limit. I'd try to be down in 30 minutes.

Turn off as much electrical stuff as possible and operate with one nav-com. If the controller insists, you might use the transponder, but it should be easy to work a deal to have the transponder operating only when and if absolutely necessary for the controller to handle the flight in an expeditious manner.

In turning things off, be very methodical. If, for example, you missed the rotating beacon and the pitot heat, you'd have left *on* the two items of highest electrical drain.

It goes without saying that a complete engine failure is an honest IFR emergency. This is true whether the airplane is a single or a twin.

---

## ENGINE LOSS

In a single, there's not a lot of decision-making when the power fails completely. There are many things to do, though. First and foremost is to get the engine going again if possible.

The chances are good that it stopped because of something the pilot did, and unless this happened to be the act of using all the fuel, the next step is to right the wrong.

In case of fuel-system mismanagement, it takes time to get an engine going again. This is why tank switching, management, should be done at a noncritical time in fight. That last switch before an approach should be back at the time the descent was started from cruising level, for example. There a mistake can be handled easily. Switch tanks incorrectly just before crossing the outer marker inbound, though, and the mistake will be difficult to handle.

In the unlikely event an engine failure is not pilot-induced, there are still many things that might bring life back to the power plant. If applicable, the carburetor heat should have been pulled at the first sign of power loss. Move the magneto switch from "Both" to "Left" to "Right." If it runs, or runs better, on "Left" or "Right," leave the switch in that position. Turn on the auxiliary fuel pump if there is one. Enrich the mixture. Check the primer as locked, if applicable. Switch tanks even if there is gas in the one selected. Change anything that has a relationship to the engine. True mechanical breakages are rare. Engine failures are more likely related to fuel or ignition, and these things have handles and switches to set and reset. Try everything.

All the while, be active with the radio, but do not change to 121.5 the instant something happens, and don't squawk 7700. If IFR, you are likely in contact with a controller, and if there is radar coverage in the area you would be in radar contact. Just stay on the assigned frequency, tell the controller about the unhappy event, and let him get on with clearing any aircraft that might be IFR at a lower altitude. If he wants a change in frequency or transponder code, he'll say so.

The controller might also be able to give information on a nearby airport. There have been successful power-off IFR approaches (I know—my father made one once), and you'd hate to miss the chance to pull that off if there was an airport around.

If there is no airport within gliding range, spend the time devising the best gliding plan for the terrain. If there are ridges, glide parallel to them. If there is an Interstate highway, glide in that direction. If you are over a lake, glide toward shore. This does assume that you have a visual chart

on board, and I think it is unwise to fly without at least
World Aeronautical Charts when operating IFR. If you don't
have the charts, perhaps the controller could tell you something about the lay of the land, perhaps not. It never hurts to
ask. Too, it is good to be aware of the surface wind so that
your impromptu arrival can be into the wind.

Unless it's zero-zero, you are going to break out of the
clouds before reaching the ground. From there on in, it is
the same as landing VFR.

If there is any doubt at all about how much ceiling there
will be, the airplane should be configured for the softest possible touchdown early in the proceedings. This varies from
airplane to airplane, and it is a good idea to know the
gear/flap/speed arrangement that results in the lowest possible
combination of forward speed and sink rate while still allow-
ing good aircraft control. Try it out in the practice area some
day. Note too that this would be a time when shoulder harnesses would appear to be the most valuable equipment in
the airplane. Given anything other than the most impossible
circumstances, shoulder harnesses will help make such a
landing nothing more than a bruising inconvenience.

## TWIN

A power failure when IFR in a twin involves as much as or
more than a power failure in a single. There are more decisions to make, and a bad decision in a twin will probably be a
lot more costly than a bad decision in a single.

To begin, the drill in a twin is much the same as in a single. Communicate the problem while trying to get the engine running again. If it won't run, feather the prop. Then take stock of the airplane's capability.

If you're flying in an area where the minimum en route altitude is higher than the airplane's service ceiling, the outcome of an engine failure might be much the same as in a single. The task is thus the same. Arrange things in the most advantageous manner, considering the terrain. The operating engine can be used to minimize the descent rate, but watch the airspeed. If you get a bit slow in a conventional twin with one engine operating and the other one shut down, the loss of control that could follow is likely much more serious than any forced landing that you have ever imagined.

A more likely event in a twin finds the airplane capable of making it to an airport after one engine fails. All the pilot must do is take advantage of the airplane's capability. This, however, can be a difficult task. The "i" must be dotted and the "t" must be crossed with a precision that is unknown to many general aviation pilots. That's why the twin has relatively more engine-failure-related fatal accidents than the single. We'll explore proficiency in Chapter 15, and will confine the discussion here to the decisions.

Once it is decided that the airplane is maintaining altitude, the next step is to choose the airport for an arrival. Snap judgment might well suggest the closest airport, but this isn't always the best place to go. Weather, field length, and terrain are all important considerations.

For example, rather than shooting a VOR approach to minimums at an airport with a 3,500-foot runway that just happens to be 10 miles away, I'd rather fly 100 miles on one engine to a big airport with an ILS and a bigger runway. Why? Because once you reach a certain point in the approach with a twin, the airplane demands absolute perfection. And there is a point in the approach at which you become committed to the landing. I would rather try my perfection on a precision approach, and I would rather play the game of committing the airplane to a landing at an airport with the longest possible runway. Another item: I would hesitate to shoot an engine-out approach in grungy weather to an airport without weather-reporting facilities. I'd just not want to give up the haven of altitude and begin an approach unless I had a pretty good idea that the weather would allow it to be flown through to completion. After one engine fails it would be an unlucky day indeed if the other one should fail, so I'd be willing to fly a long time for the best possible deal on an approach. And I would be much more at ease if I had gone to the trouble of maintaining proficiency in instrument approaches to minimums with one engine out.

Once the approach begins, the decisions must be based on closing doors behind the twin with an engine out. Specifically, when the decision is made to descend below 500 feet above the ground, or to extend full flaps, the act should be accepted as an irrevocable decision to land the airplane. Once those things are done in the average light twin, the available performance is too marginal to consider anything other than a landing.

## ——————— WEATHER EMERGENCIES ———————

There are what might be considered emergencies in relation to weather. If, for example, a pilot bumbles into a thunderstorm, he or she may consider that an emergency. It isn't one in the true sense of the word, because there is no applicable immediate action other than remembering the principles and doing the best possible job of keeping the airplane upright. The storm will pass, and unless the airplane sustains structural damage, the flight after the storm will return to a routine operation as the pilot's knees slowly cease knocking.

Ice can be considered a weather emergency, and has already been covered.

The other weather emergency that we might someday face has to do with blown forecasts. What do we do if the destination goes to zero-zero, the alternate goes to zero-zero, and nothing else within fuel range has landing minimums? I think it goes without saying that a pilot who backs into such a corner probably made some unwise weather decisions based on wishful thinking, but once there you still have every right to try to extricate yourself without damage.

It is extremely hazardous to even think about an emergency landing in below-minimum conditions unless the airport has a full ILS system. With the ILS, things look up, and a sharp aviator can put an airplane on an ILS runway in the foggiest weather if the airplane is flown with absolute precision.

In making an ILS approach in below-minimum conditions, I'd configure the airplane for landing at the outer marker. I'd put the gear down if in a retractable and set the flaps at the takeoff or approach position. If no takeoff or approach flap position is specified, I'd go for a partial flap setting that reduces stalling speed but still leaves the airplane in a more or less level flight attitude on the approach. For an approach speed, I'd select a value 40 to 50 percent above the stalling speed. Then I'd fly the ILS in the normal manner until reaching the decision height. At DH, I would leave the descent as it had been while tracking the glideslope. Power and aircraft attitude would remain exactly unchanged. I'd then disregard the glideslope needle and concentrate on the localizer. If the localizer needle is kept in the center, the aircraft will fly to the center of the runway. The glideslope will usually lead the airplane lower than 200 feet with precision, but you really don't need it. If the airplane has been tracking toward the runway in a vertical sense, and if it is on the glideslope at the decision height, a continuation of the status quo will lead to the runway. The glideslope leads the airplane to a point 1,000 feet down the runway, so there is some margin. There's less margin in the left-right sense. That is why the localizer needs all the attention.

It's seldom so zero-zero that you won't see any runway and won't be able to flare. But if it were that bad, I might make a very slight nose-up adjustment in attitude at about 50 feet. That's all. Then I'd work like the very devil on keeping the localizer needle centered and wait for the wheels to contact the runway.

## GCA

Better yet, if I were to be backed into a corner with nothing but zero-zero around for an approach, I'd go to a military base and ask for a precision ground-controlled approach. I had a friend who once did this in a light twin. He was deluged with paperwork after his landing, but he strongly felt that it was worth every bit of it. With only a few gallons of fuel on board, had he not gotten to the proper place and made the approach the first time, he'd have been up a tree. Or in a tree.

---

## COOL HEAD

Regardless of the type of situation that might come up, the one sure thing is that a steady hand and a cool head will help save the day. It takes discipline to be calm and collected when things start going to the devil in a handbasket, but if we recognize that calm and collected is what wins the battle, the motivation should be there. And remember that time spent reflecting on how you might handle all conceivable situations is a good investment. If you have thought it through at least once in advance, the actual event will be at least one word removed from a true emergency. That word is "unforeseen."

# 12

# THE SYSTEM

I like to think of the "system" as the direct government involvement with flying. Some consider it to be just the traffic control business, but there is enough interrelationship for us to consider the total. And from the beginning do acknowledge that the system is *not* something that works automatically or that protects. The system is something that pilots must make work for themselves. What we get out of it is in direct proportion to what we put into it.

In examining the government's relationship with the IFR pilot, look first at the rules on qualifying to be an IFR pilot.

We don't have to fly actual instruments, in cloud, to get the rating. Instrument instructors may never have flown in cloud, and I'd be willing to bet that there are FAA inspectors who have never conducted IFR operations in a light airplane and who would not willingly do so. A pilot who goes through and gets an instrument rating in such sterile conditions, from people without actual experience, had best know that his or her rating is only an indication of success at jumping through a government-prescribed hoop. Ability and knowledge have to expand far beyond that in hand and mind at the time a rating is acquired under those conditions. The pilot need only recognize this to be okay. Ideally, it would be recognized early in the training process and the pilot could switch to a school or an instructor that teaches instrument flying instead of test-passing. But if a pilot winds up with a "dry" instrument rating, meaning that the pilot hasn't actually been in cloud, there is nothing to keep that pilot from hiring another instructor for the wetting of the rating.

The government's reluctance to prescribe the actual experience in training is as it should be. The rules are minimums. We have to be wise enough to know what the word "minimum" means. Look it up in your dictionary if you need a refresher.

The minimum relationship is there as we move to the rules governing proficiency. Note that the only requirement is for so many hours of instrument flying in the recent past plus a prescribed number of approaches. We desperately need a better proficiency program than the one required by law, and this will be fully covered in Chapter 15.

Look next at the rules governing hardware: the airplane and equipment. Minimum, and very few general aviation pilots fly IFR with anything approaching the minimum in equipment. Some of the requirements—altimeter and transponder checks, for example—seem of more nuisance value than anything else, but the government must play its game.

## ALTITUDES

When we come to altitudes, we must really come to terms with the meaning of the word "minimums." Some feel that government-prescribed altitude minimums are a decree of how we should fly. That is not the case. Instead think of minimums as numbers defining the lowest possible altitude at which we can fly without hurting. The margin for error in minimum altitudes is slim, especially on approach, and we cannot fly with the thought of being plus or minus a hundred or so feet. It had best be all plus.

## ICE AND STORMS

The rules don't help much when it comes to bad things like thunderstorms and ice. There is no rule against flying into a thunderstorm, and there is no specific rule about flying into icing conditions in light airplane IFR that is not for hire. The lack of regulation here is good, because any law prohibiting flight in relation to dynamic weather situations is impractical at best. When the FAA does try to write a rule on ice, the rule becomes ambiguous. For example, the air-taxi regulations prohibit flight into known or forecast icing conditions (unless the airplane is equipped with de-icing equipment) and then hedge by saying that if current reports and briefing information indicate the ice won't be there, it is okay to fly on. The buck is clearly passed to the pilot. That's where it belongs, and that's the only place it can be handled.

The pilot must only realize that no outside source, no mystical "they," no system, no regulation, offers any protection against the elements.

To this point we've talked about regulations. There is no shortage of those, and while we can find some guidance in them, common sense and a strong feeling of responsibility will often carry the day better than any rule.

For example, a new flight instructor asked me what I would do if I had a communications failure and then encountered potentially debilitating icing conditions at the altitude that I would be expected to fly according to prescribed lost-communications procedures. No rule covers that. I'd squawk 7700, the emergency code on the transponder, and go about the business of extricating the airplane from the icing condition in the quickest and most effective way that I could devise. In fairness to others, I would try to imagine where other IFR traffic might be operating and would do everything possible to avoid any possible conflict. I sure wouldn't look to any rulebook for guidance.

## THE PEOPLE

When we set out to fly an actual IFR flight, the point of first system contact is usually the flight service station. This is an important part of the general aviation pilot's IFR system, because it is where we start working with people. It is also the prime point of information. Again remember that the system is not automatic. Even with government people involved, we have to make it work for us to get maximum benefit.

Some pilots tend to put themselves at the mercy of the FSS briefer, as if to leave the go/no-go decision in his or her hands. This is bad, because if a pilot doesn't have the intelligence to obtain information, interpret it, and make a sound decision on a flight, then the pilot's training is inadequate and his or her fanny is in danger. Don't forget that an IFR pilot should, for self-protection, strive to know more about the effects of weather on the airplane than any briefer at a flight service station might know.

## AIR TRAFFIC CONTROL

So far as human relationships go, the IFR pilot spends far more time in the company of the air traffic controller than with any other person we deal with in flying. In a lifetime, an active pilot might spend more time in conversation with air traffic controllers than with his or her own kids. Certainly we spend more time with the controllers than with an instructor, the inspector, or flight service station specialists.

The proper relationship with the air traffic controller should not be master/slave (in either direction). And the old question asked of a controller by an airline captain must be qualified. The captain said: "Am I up here because you are down there, or are you down there becuse I am up here?" In truth, it works both ways. If we didn't fly IFR, they wouldn't have a job, but if they weren't doing their job, an IFR flight wouldn't work very well.

The part about making the system work for yourself is very pertinent when dealing with air traffic control. The controller makes it work for himself—the job, pay, and working conditions are quite good, in fact—and the controller will make the overall traffic control system work for everyone on general terms. But the individual pilot must find the proper piece of airspace in which to fly, file a flight plan outlining the use of that airspace, operate the airplane per the clearance, and request any changes deemed advisable or necessary once en route.

It is very important not to feel a slave to the system in time of need. For example, if you need to fly at a low altitude— the published minimum en route altitude shown on the chart, for example—the controller might balk. If there is controller reluctance to this, it might be caused by the fact that radio or radar coverage isn't good at low altitudes and controllers do not like to lose contact with airplanes. But there is a way to do it, and if you need that low altitude and make the request clearly, it should be honored. The controller is paid to do any extra work that might result from your being out of contact for a while, and indeed, there is also extra work for the pilot in flying low.

---

### QUESTION

If a traffic controller will not approve a request, I feel that it is okay to ask one question. It is: "Is my request denied because of traffic or for procedural reasons?" If it is denied for traffic reasons, that's okay. Nothing more to say. If the request is denied for procedural reasons, then it is proper to question the wisdom of the procedure by telephone, after landing. It is sometimes tempting to argue while in flight, but that is not what the radio frequencies are for. Fussing should be reserved for other times and certainly the questions should be raised and the issues thoroughly explored any time a pilot feels that the air traffic control system is being run for the convenience of the controllers instead of for the users. The air traffic control system does get into some dumb and wasteful procedures at times, and if pilots don't squawk and demand more efficiency it'll surely get worse before it gets better. But do the squawking on the phone or in person.

## OTHER SIDE

Pilots are certainly not perfect; in fact, controllers do their work in a more uniform and skillful manner than do general aviation pilots. If we are to pick at their system, we need to work to understand their problems and to correct our own areas of need. One way to start is by trying to understand the basics of the air traffic control task.

An IFR pilot who has not visited both an air route traffic control center and a terminal radar control room is lacking the important overview of the system that is necessary to keep from being a misfit. Just watching the handling of a flight helps to shed light on many areas of misunderstanding. For example, you can't really get a feel for the controller's problem at a big and busy airport until you watch the line of blips on final, three miles apart, all fed to that line from various fixes in the terminal area. Watch that for long enough and a general aviation airplane will show up; the controller is likely to ask a question: "What will be your speed on final?"

Speed is an important thing to the controller, and we should understand that the more nearly we conform to the flow of traffic, the better. One rainy day at Washington National I heard a Bonanza pilot illustrate how not to do it by answering: "Eighty knots." He didn't understand the problem. The big jets are going to be flying at about 120 knots when inside the marker, and fitting an 80-knotter into the string is difficult. The controller had no choice but to make extra room. The Bonanza pilot could just as well have flown his final at 120 knots, smoothing both the flow and the general aviation pilots' image.

In watching controllers do their work, the two-dimensional nature of the air traffic control system is perhaps the most impressive thing to a pilot. The airplanes all appear on one flat radar screen. The controller has to look at and interpret the altitude readout to know that the airplanes are properly separated. Without the automatic altitude information, the controller must ask the pilot for the information.

As we fly along VFR, we might not see a lot of traffic, because we have vertical as well as horizontal separation. But the scope can indeed appear crowded and hectic to the controller at the same time.

The way the controller sees it is okay for his purposes, because the task is to keep airplanes separated as they move to and from airports, and in the crucial beginning and end flying *is* two-dimensional. All airplanes start and stop on the ground.

## THE RELATIONSHIP

There have been and will always be a lot of differences of opinion between pilots and controllers. The gap has widened in recent times, for several reasons. Foremost is the fact that an ever-smaller percentage of controllers are active pilots. I think a pilot working at the job of controlling traffic naturally puts more feeling into the task, and has a better understanding of the pilot's problem. When a person calls and requests a different altitude because of ice or turbulence, a controller-pilot might well have been in a similar situation, and knows how eager one can be to have a change in altitude. Likewise, one who has flown in the canyons of cumulus knows how helpful it is to have word on any weather return from the traffic control scope.

The controller who doesn't fly is more likely to look at the job as a task that isn't even aviation-related. Certainly if you visit a center or approach control facility you'll see that the atmosphere doesn't have a strong aeronautical flavor. Those people might just as well be solving mathematical problems with a computer, running a TV station, or playing an electronic TV game.

We are probably going to have to get used to an even colder and less personal attitude on the part of traffic controllers. As aviation grows, the fraternity aspects of it are fading away. And as air traffic controllers become less personally related to aviation, they also seem to become more sensitive. This contributes to a rift between pilots and controllers. Pilots can freely criticize other pilots, but some controllers rail at any critique of their efforts by pilots.

Finally, a sense of ill will between controllers and the FAA doesn't help the pilot/controller relationship at all. This started some years ago when the controllers struck (though they didn't call it a strike), and it has grown over the years. Some now feel that the FAA has gone too far in placating the controllers, but the controllers hardly agree with that. Whatever, labor-management strife is likely to continue, resulting in a less enthusiastic attitude on the part of air traffic controllers.

None of this is offered as criticism of anyone. It is just a fact that traffic control is becoming a less personal business, and we might as well know why this is so.

## THE MIGHTY COMPUTER

Moving to the inevitable computer, fitting into the scheme of things in our air traffic control system program is important. If an airport has published standard instrument departures and/or arrivals, the flight plan had best be based on the use of these. If there are preferred routes published for the trip, the flight plan should be based on the published routes. This makes us feel subservient to the computer, but you can take some long delays while they fool with a flight plan that doesn't fit into the programmed way to go. If the program is ridiculous, as some are, take it up later with the powers that be at the traffic control facility involved.

Using general aviation airports for IFR operations is at times complicated by the fact that there is no direct communication with air traffic control when you are on the ground awaiting departure. Where there is a will there is a way, though, and if there is no radio contact, the telephone becomes one means of communication. Even that can have its perplexing moments. I remember leaving a small field early one Sunday morning, before the office opened. I had an IFR on file, and the weather was below VFR minimums, so it wouldn't be possible to depart and then get a clearance when in radio contact. No place to call at the airport for sure, but then I remembered a roadside phone booth a couple of miles back. I drove there, called the approach control facility covering the airspace, and got a clearance with a void time. It worked fine, and I was soon off.

Do be wary on IFR departures from VFR airports. If an airport with an approved approach has any necessity for special departure procedures to avoid terrain or obstructions, this will be noted on the chart. Leaving a VFR airport under IFR conditions, you must study the situation yourself and make certain the climb path and gradient will provide proper clearances. The system offers no protection in such a case.

The air traffic control system was designed for the air carriers, but the byproduct that has become our part of the system is still quite good and workable. But, again, we must make it work for ourselves. The local airport won't have an instrument approach unless someone requests that one be approved. Your IFR flight plan doesn't go on file until you make the call. The decision on weather is yours. Once you're en route, the altitude changes and deviations in flight path necessary for comfort and safety are of your choosing. The friends (or enemies) made as we fit into the flow of traffic are products of individual actions. And our relationship with the people of the "system" is made good or bad on an individual basis. Be positive but smile when you speak, and things should go more smoothly.

# THE MACHINES

The first airplane that I used for instrument flying was a Piper Pacer. It was a basic airplane that taught me a lot of lessons about the machine versus the IFR environment.

My Pacer flying was done in the mid-fifties, and while the airplane was well equipped for the day, it was a little short by more modern standards. As a primary bit of avionic gear, I had a Narco Omnigator which provided VOR and localizer plus VHF communications and a marker-beacon receiver. An ADF was fitted, and in three years I went through as many different radios in a space on the right side of the panel. First there was a low-frequency receiver and VHF transmitter. But low-frequency ranges were on the way out, and I replaced it with another Narco, a Superhomer, that gave me a backup VOR but with accuracy that was not really adequate for IFR flying. That gave way to a crystal-controlled

60-channel transceiver, to improve on the ability to communicate. The Pacer also had a wing-leveler autopilot.

Except in the rare instances when a radio would malfunction, I never felt short. I could navigate and talk—enough for the time—and the wing-leveler would handle that chore when I needed to look at a map. It was before the advent of DME and transponders, so those items were not even missed. And to my knowledge there was not a glideslope receiver available for smaller general aviation aircraft, so I had no real basis for wanting in that area. In retrospect, it was the lack of a glideslope that put the most severe limitation on the IFR use of the Pacer.

Minimum descent altitudes (they were just plain minimums then) for VOR, or for localizer approaches, were usually 400 to 500 feet above the ground, with required visibility values from a low of half a mile on some localizer approaches up to a norm of a mile for most non-precision approaches.

The weather seems to know about those minimums, and the majority of instrument approaches seem either in conditions comfortably above the 500-and-1 specified for many non-precision approaches, or between that value and the 200-and-½ we most often see listed as the acceptable decision height and visibility values for a full ILS approach.

Pitting an airplane of the Pacer's performance and range—100 to 110 knots, five to six hours of fuel—against a basic 500-and-1 weather requirement often created sticky situations. This was especially true when there was a headwind to contend with. The thing that I learned quickly was that endurance is extremely important in an IFR airplane, especially in a slow IFR airplane, and that a glideslope receiver is a very desirable item because of the lower minimums it affords.

-------------------------------- **ENDURANCE** --------------------------------

Endurance is of critical importance, because it takes a lot of it to create range, especially with a headwind. The weather is only interested in the progress you make in moving through the features on the map. If you have arrived at Wichita to find zero-zero, and the closest sure shot for good weather is Denver, a Learjet's hour against westerlies can become a Skyhawk's four and a half hours.

For an example of how endurance becomes more critical the slower the airplane goes, look at two different airplanes with five hours' endurance: A 40-knot headwind cuts the effective range of a 120-knot airplane down to two-thirds of its original value while slicing the 140-knot airplane down to 71.4 percent of what it started with. A headwind of given value always cuts a greater percentage of the range of a slower airplane than of a faster airplane.

This effect can be more pronounced when you subtract the alternate and reserve requirements for IFR. Given 45 minutes for the law plus 80 nautical miles to fly on to an alternate, a five-hour and 120-knot airplane with 40 on the nose gives you but three plus 15 to match against en route distance. Thus, the available range for flight planning is 260 nautical miles. That is not too generous. The departure point, destination, and alternate can often be under the influence of the same weather system on hops that short.

The 140-knot airplane, by comparison, still has 345 nautical miles range with the same reserves with a 40-knot head wind. Eighty-five miles can often make a big difference.

Add the higher weather minimums required in a no-glideslope airplane to the range restrictions with a headwind in a slower airplane and you can see why the ability to shoot lower approaches and choose alternates with worse forecasts is worth a lot.

## WEST VIRGINIA

Endurance combined with a little extra speed can make a more dramatic difference than is immediately apparent in some hypothetical IFR trips. A good example relates to the mountains of West Virginia. This is not a large state, but it is often one of the more difficult ones in the nation to cross westbound, especially in the wintertime.

The mountains affect the weather a good 50 miles to the lee side, to the east in the normal winter circulation, and you start noticing this some miles before reaching Martinsburg, West Virginia. Up- and downdrafts start appearing, and cloud tops tend to become higher. From the point where you start encountering the effect of the mountains to where it starts ending is 200 nautical miles, minimum. This means that with the usual headwind, a slower airplane might be testing its maximum IFR range just in passing through this effect.

Often the choice is between landing and refueling just before starting across, or facing the possibility of not having adequate fuel to make it across. The latter isn't without op-tions—there are places to stop—but the weather is usually worse in there. If there's a chance of ice anywhere, it'll be in cloud over the mountains, and the turbulence clouds gener-ate can be very worthy of note. In a 140-knot airplane with six hours' gas, I've never failed to make it from the New York–Philadelphia area to at least Charleston, West Virginia. In a 120-knot airplane, I've had some rough trips with an extra stop.

### UP AND OVER

Climbing and ceiling go hand in hand with speed, because the extra power used to make an airplane go faster can as well be applied to climbing. The basic four-place airplanes all climb at between 600 and 700 feet per minute and hit the ceiling at 13,000 feet. The 140-knotters climb almost half again as well, and most have ceilings above 15,000 feet. In the wintertime, ceiling is often the only tool we have to use in icing conditions, and an airplane that is unable to get to at least 15,000 feet can be at quite a disadvantage.

An example comes to mind. The flight was across West Virginia; luckily it was eastbound, with the wind.

Tops were reported as high as 17,000 feet, and conditions strongly suggested that the only way to go across would be on top. The low-level weather was a combination of wind-whipped turbulence and icy clouds.

I was flying my Cardinal RG this day, and had oxygen bottles along. In fact, I don't like to fly wintertime IFR without oxygen, because it can often be a go/no-go item. You sure can't fly at 17,000 feet without it, and to me it would have been no go this day if I could not have gone that high.

It took some coaxing to get the airplane to 17,000 feet, but it finally made it. The climb followed a takeoff at Lexington, Kentucky, and I was able to avoid most clouds (and thus most ice) in the climb and reach an on-top condition quite easily. Tops were about 11,000 feet over Lexington, with the height increasing as I moved eastward. It was a rather clear illustration of what the mountains do to cloud tops. The highest general tops built to 15,000, with a lot of buildups to over 17,000. My trip across was smooth and serene, with a few deviations around buildups. A lower trip would have been a bad one.

The altitude capability of the airplane made this trip possible. This is why turbocharged singles are appealing IFR airplanes. Had I had one of those this day, I'd have probably moved on up to 21,000 and enjoyed a blistering tail wind up there plus the best available true airspeeds. As proud as I was of my RG, it was out of ideas at 17,000 feet.

## WHICH?

We have noted speed, endurance, climb, and ceiling as performance parameters that can have a very definite effect on instrument flying. It is interesting to try to put the four in order. Which one means the most?

Certainly speed is worth nothing in light airplanes without endurance, if for no other reason than the reserve requirements. Even at a cruise speed of 200 knots, a three-hour airplane becomes restricted when you take off 45 minutes for reserve and, say, an average of another 45 for a trip to an alternate. It becomes an hour-and-a-half airplane, and a nervous one at that.

Of those first two items, endurance is more important to me than speed except where the speed advantage would be great. For example, given the choice of a three-hour 140-knot airplane and a six-hour 120-knot airplane, I'd take the latter unless average trip length happened to be less than 150 nautical miles. And even then the speed advantage would amount to a very few minutes, and the endurance might be nice to have for other things. Even with a choice between six hours at 120 and four hours at 140, I'd take the six hours. If the choice were six hours at 120 or four hours at 160, I'd begin to waver.

Rate of climb does a lot of things for an airplane in instrument flying. I remember a question that was raised while I was leaving Dallas one warm winter day in my Skyhawk. A colleague was flying the airplane, which was loaded to gross. As we struggled upward at 500 or less feet per minute, the person flying remarked that a larger engine would surely be nice in the airplane, if for no other reason than to improve the anemic climb rate. "Why," he said, "if you encountered the least ice and the tops were as low as 5,000 or 6,000 feet, you might never make it on top when climbing like this."

There are two ways to buy climb capability. One is with raw horsepower, the other is with turbocharging. When gross-weight climb figures are examined, the best sea-level climb we generally see for single-engine airplanes is in the neighborhood of 1,000 feet per minute, plus or minus some. Add turbocharging to a single and it won't climb any better at sea level but it will maintain rate of climb as it goes up. Couple this with a high service ceiling and increasing true airspeeds aloft and it is obvious that the turbocharger can offer answers to IFR questions. The combination of an aerodynamically clean airframe and an efficient and reliable turbocharging system is hard to beat.

Buying a twin is another way to get more rate of climb. If a twin will climb at all on one engine, that means that the power of the other engine can be devoted to extra rate of climb when you are operating with both engines. The results are comparatively spectacular. Most light twins go up half again as well as the best singles, or even better. The rub comes at the gas pump, and with fuel efficiency becoming a big thing, carrying a lot of extra horsepower around for the sole purpose of being able to climb better might not be an enduring proposition.

Most IFR users find an airplane that climbs somewhere near 1,000 feet per minute to be adequate. Climb rates below that involve compromise and might cause an occasional cancellation due to low rate of climb or correspondingly low service ceiling.

When comparing high initial rate of climb, as with an unturbocharged twin, and high service ceiling, as with a turbocharged single, the practical IFR aviator would probably opt for the turbocharged single. The value there would have to be balanced against the missions flown. In the west, an airplane simply is not an IFR airplane unless it is turbocharged. In the east, wintertime trips can often beg for turbocharging to avoid ice or to take advantage of strong westerlies aloft on eastbound trips. In the summertime turbocharging can keep the climb rate peppy up to comfortable IFR altitudes, and can often make possible a flight on top of haze and murk. Without it you are down where you can't see the cumulus and cumulonimbus imbedded in the smaze until the last minute; with it you can be at Flight Level 200, surveying the scene without restriction to visibility.

------------------------------ *ON TOP* ------------------------------

There is one operational consideration to watch with turbocharging. I found it in a Turbo Centurion as I moved smoothly toward my destination one day at Flight Level 240 (24,000 feet). Cloud tops were about 22,000, they were uniform, and I had not given a lot of thought to pending weather problems as I neared the destination. Then the traffic controller told of a line of weather at 12 o'clock, 30 miles, and cleared me for descent. I sure couldn't see it from on top, but it was quite wet and bumpy down in the clouds. I wished mightily for airborne weather radar to help me find a smoother path.

When flying high (meaning in the twenties—high for all but the jet set), we do fly up over a lot of weather, and our thinking needs to be adjusted for the altitude. When flying on top of clouds at or below 10,000 feet, we pretty well know that passing through the clouds on the letdown should not involve any extraordinary turbulence. When flying on top at 24,000 feet, though, there exists the potential for a lot more weather and a lot more turbulence in the letdown. That doesn't detract from turbocharging, it just outlines a new consideration.

How do we wind up on the performance parameters? It is obvious that endurance is the most difficult thing to do without. Next might come ceiling, then speed, then rate of climb. Rearrange to suit your need but don't fail to evaluate each one in relation to your need when choosing an IFR airplane.

## HANDLING QUALITIES

How an airplane flies is very important to the IFR aviator, and preference here is usually based more on the pilot than on any envisioned mission. If the pilot is a tiger about staying proficient, the aerodynamically slickest airplane is the best deal, because it is the most efficient. If a pilot feels that he or she won't be doing a lot of IFR flying and the missions are generally short, then an airplane that is slower might be more forgiving of error and the better deal. An airplane in the class of the Piper Warrior, for example, is very good for a pilot who wants a nice stable platform that doesn't rush things. Remember, the faster you fly the faster you have to think. An IFR flight between here and there involves about the same number of tasks in a Learjet and a Warrior, but there is less time for the work in a Learjet.

In considering the fine points—control forces, trim changes, lateral and longitudinal stability—it always seems best to fly an airplane and see if it fits your hand and bottom.

Some pilots like light elevator forces, some like heavy elevator forces, and there are debating points for both sides. It is possible to find a correlation between elevator forces and instances of in-flight failure—the lighter the required elevator forces, the more involvement in structural failure—but this is logical, and perhaps it does not even bear on the overall safety record. Most in-flight failures come after a pilot has lost control of the airplane. This means high airspeed and a gangbusters rate of descent. The pilot likely perceives the problem and attempts a recovery. The lighter the required elevator forces, the less effort the pilot has to expend to break the airframe in a recovery attempt. The reason that this might not bear on the total accident picture is that the outcome could be the same with light or heavy elevator forces: The airplane would be lost. The difference would be in whether the airplane hits the ground in one piece or more than one piece.

The airplane with lighter elevator forces is perhaps nicer to fly much of the time, but in a loss of control it could be more demanding.

The question: Do you want an airplane with elevators that are delightfully responsive to the touch, but that might bite if you let it get away from you? Or would you rather have an airplane that is trucklike but that could be a little easier on you in an extremely unlikely moment of upset? Six of one and a half dozen of the other. Fly them all and make your choice. Do note that at times the airplane with heavy elevators is the easier one in which to hold altitude precisely.

## ROLL

Even if the elevators are leaden, we can use the trim and still fly with slight pressures. This isn't so with lateral control. If the roll rates are slow and the required aileron forces high, an airplane puts you to work and keeps you at work all the way through a flight. Some pilots would rather wrestle with an airplane than fly it; they no doubt like heavy ailerons and slow roll rates. I like them a little lighter, but an airplane that is exceptionally quick in roll can be quite distracting in instrument flying. Again, fly and take your choice. The one that finally strikes you as best will probably have some balance to it—the required lateral and longitudinal forces will seem equal—and will probably reflect your personality to some extent: deliberate and middle-of-the-road or quick and dynamic.

## RIDE QUALITY

An airplane's response to turbulence is quite important in instrument flying, because this combines with handling qualities to put the pilot's ability to the test in turbulence.

Lateral stability in turbulence is by far the most important, because a lateral upset almost always precedes a longitudinal upset in a light airplane. Directional stability is also important, as is proven by the amount of money spent for yaw damper systems on more exotic airplanes. Whether any correlation could ever be drawn between yaw stability and loss of control, I know not. I can only say that I have never seen an airplane prone to yaw in turbulence develop an unusual desire to do anything other than just yaw.

Dutch roll, the lateral/directional couple, is tough to deal with in turbulence. The airplane rolls *and* yaws from side to side (the wing tip makes circles on the horizon); fortunately, no current production airplanes have an extremely strong dutch-roll tendency.

The size of the tail surfaces on an airplane often tells a tale of how it will ride in turbulence as well as how it will fly. Bigger is usually better. In examining numbers, wing loading is the one most directly related to riding qualities. The higher the wing loading, the better the ride. Other things do indeed affect bounce qualities, but wing loading is the primary number to look at.

Certainly, in picking an IFR mount it is good to fly the airplane in turbulence and sample its ride and handling qualities. An airplane that turns you on otherwise might seem altogether too busy in turbulence, and the time to learn that is before rather than after.

## MANEUVERING SPEED

Maneuvering speed is worthy of note because this is the speed at which you'll be flying when the going gets rough. If maneuvering speed is quite a bit below the normal indicated airspeed at cruise, it will mean slowing down just that much in turbulence. Some see a low maneuvering speed as an all-bad thing, but it can have its advantages. You might have to slow down in turbulence, but a low maneuvering speed also means a low stalling speed, which in turn is a better deal should it ever become necessary to land the airplane on some surface other than a runway. (Maneuvering speed is a product of stalling speed and the airplane's limit-load factor.)

The lower maneuvering speed often also means lighter wing loading, though, thus more response to turbulence and a bumpier ride. Perhaps the best of all worlds is having an airplane with a high flaps-up stalling speed, a corresponding high maneuvering speed, and an effective flap system that lowers the stalling speed as much as possible for landing.

## GETTING DOWN

Another important speed-related item is an airplane's descent capability. The most demanding IFR descent situation is where the desire is to lose altitude as rapidly as possible in an area of turbulence that necessitates flying at maneuvering speed. Keeping the engine(s) warm during such a descent is also a consideration, and many airplanes offer more questions than answers in such a situation. If there is no way to descend at a rate of 1,000 feet per minute at maneuvering speed with the engine developing enough power to stay warm, there will be awkward moments or times of rapid engine cooling. The latter can have a very detrimental effect on engine life. Note also that in almost all airplanes the use of approach flaps would not be allowed in this descent situation, because the use of flaps lowers the limit-load factor of the airplane. That is not a desirable thing to do in turbulence.

When flying a retractable, the landing gear often serves well as a speed brake in a descent. Do note the relationship between the maximum allowable landing-gear speeds and maneuvering speed, though. The ideal situation is with the gear speed at or above maneuvering speed.

Being able to control speed in turbulence is important and one related factor is worth noting again and again. Generally, the turbulent air speed should be reduced when the airplane is flown at lighter weights. I know this is contrary to what some think is logical, and it always raises arguments. But it is based on sound principles and fact. For one thing, the spanwise distribution of weight changes with fuel burnoff, resulting in a greater concentration of weight in the center and less relieving weight in the wings. This increases bending loads. Too, the wing loading is lower at lighter weights, so the airplane will experience greater accelerations in any given gust, putting greater stresses on the total airframe. Some misinterpret all this as suggesting that the airplane is stronger when flown at heavier weights. That is not necessarily true. Basically, the whole airplane is a known and established quantity at its gross weight. If the total machine's ability to withstand a vertical gust of a given strength is to be maintained at weights lighter than gross, the speed must be reduced to help manage the product of that gust.

## FUEL INJECTION

A powerplant feature worth pondering is fuel injection. This often offers the disadvantage of hard starting when hot and poor idling qualities in hot weather, but all the minus points are related to that limited time, and fuel injection is very nice when flying IFR.

There is no carburetor heat to fool with, and most systems either have an automatic alternate air source or an air supply that is not subject to impact ice or clogging in heavy snow. Too, fuel-injection engines have better mixture distribution and can be leaned more precisely. All around, it is a better deal for an IFR airplane.

## THE PANEL

Moving from the basics of the machine to the inside, we find an almost unlimited selection of things to do with and for an airplane. The first step is to establish a basic IFR requirement and elaborate from that point as finances allow.

A basic avionics package might consist of two contemporary nav-com radios or two separate com and two separate nav radios, plus glideslope, ADF, transponder, encoding altimeter, audio selector, and a marker-beacon receiver.

Before expanding that package, I'd next add some special IFR items that often are not included in production aircraft. Static wicks are a necessity to help get rid of precipitation static, and a better navigational radio antenna than the standard cat's whiskers is strongly advisable. Beside improving reception and accuracy, one of the good antennas will minimize the effects of precipitation static on the nav radios.

Next consider the warning systems in the airplane, especially if it is single-engine. Is there a low-voltage light that would quickly alert you to an alternator failure? If not, add one. Relying on the eye to catch a needle showing a discharge as the only available indication of alternator failure is pure foolishness. What about vacuum? Vacuum-failure warning systems are not as widely available as low-voltage lights, but unless the vacuum gauge is on the flight panel and is methodically included in the scan, some obvious indication of vacuum failure should be included on the panel.

The static wicks, antennas, and warning lights are minimum items—a few hundred bucks or so will cover them—but they are things that are often left off. Too bad, because each contributes to serenity (or safety) of flight way out of proportion to cost.

A more personal item to add is a microphone switch on the control wheel. A lot of pilots resist the use of boom microphones with a wheel switch, but there are times when this can cut the workload enough to make a difficult departure or arrival noticeably easier. Or perhaps you just want to retreat into your headphones to get away from gabby passengers. Whatever, an IFR airplane should be equipped to give you this option.

A lot of us like to do something special in the way of engine instrumentation when an airplane is used for IFR. Certainly an exhaust-gas temperature gauge is a good thing to have, for more precise leaning, and the lily can be painted in this area by having an EGT system that measures the temperature for each cylinder. With this, you can check on the combustion in each one; systems are also available that give cylinder-head temperature readings from each cylinder.

Once you are past the basic improvements that seem prudent for any airplane that will be operated IFR, the choices are harder to make.

High on my list for a single is an electric directional gyro or horizontal situation indicator. Panel space permitting, this would be in addition to the vacuum-operated DG.

Most airplanes put too many eggs in the vacuum basket. I remember flying one that had a very elaborate and complete avionics system, including a flight director and autopilot, yet I was reduced to hand-flying with only a turn coordinator and the pressure instruments after an actual vacuum-pump failure. When installing or buying a full set of expensive equipment, make a careful evaluation of what will be left in case of either vacuum or alternator failure, and try to maintain as near 50 percent capability as possible after a failure of either. The best arrangement I have had featured an electric horizontal situation indicator and a vacuum DG in the panel plus the usual items, including the vacuum-driven artificial horizon. After a vacuum failure I still had an HSI and a turn-and-bank indicator with which to fly; if the alternator dropped off the line, I had a vacuum artificial horizon and DG. Either way gave me a pretty good set of instruments.

In arranging things to help handle failures with minimum fuss, acknowledge the most common problems. The most likely is an alternator failure. The reliability of these systems has not improved over the years—if anything, it has become worse—and you can almost count on some alternator problem every 500 to 1,000 hours. An alternator failure, though, is easily handled if caught promptly. There is enough juice in the battery to handle one radio and perhaps the transponder while you go to the nearest airport and land. The only requirement is that the failure be caught promptly. Thus the necessity for the low-voltage light, which comes on almost instantly and lets you start planning to maximize the energy stored in the battery.

There is also the more remote possibility of a complete electrical failure caused by a short circuit somewhere in the system. This can happen in a single or a twin, and the only guard against it is a battery-powered radio. This is something that most of us think about purchasing at one time or another, but that procrastination usually keeps out of the cockpit. Fortunately, complete electrical failures are relatively rare. It might still be wise to at least carry a battery-powered VHF portable receiver (with earphone) along. In a bad situation such a radio could become quite useful.

The vacuum failure is less likely than an electrical failure, but after a vacuum failure the affected items will cease offering usable information quickly. Given a warning here, you'll know what is happening and won't continue trying to fly (or won't let an autopilot try to fly) an instrument that is rolling over dead from lack of vacuum.

Notification of other glitches can only be a good thing for the IFR pilot. If a light quickly brings to your attention a drop in oil pressure, for example, it serves the purpose of providing as much time as possible to use in dealing with the situation.

## *AUTOPILOT*

A lot of pilots feel that an autopilot is a primary piece of IFR gear, and it can indeed be one of the friendliest things in an airplane. There are, however, considerations to autopilot use.

The basis for an autopilot might well be in the pilot's outlook on its go/no-go status. In a single or light twin, if a pilot would not conduct normal IFR operations with the autopilot inoperative, then the pilot is probably depending too much on that autopilot. And if a pilot ever lets an autopilot shoot an approach that the pilot would not attempt, said pilot is engaging in a terrible game of transistorized Russian roulette.

That aside, an autopilot is wonderful on long trips. Engage it, slide the seat back a notch, and carefully monitor the instruments as the mechanical wizard does its thing. Most autopilots fly more smoothly than humans, and you can actually improve your own technique by watching the machine's deft handling of the airplane. It works in a manner that puts a recalcitrant needle back in the center with a minimum of fuss. In the terminal area, let a good autopilot intercept the ILS and note how it nails the needles. Keep a hand lightly on the wheel and feel how smoothly it flies. But always remember that an autopilot might fail. It is likely to happen once every 500 to 1,000 hours. If you were with a real pilot who you knew would have a heart attack while flying within the next 500 to 1,000 hours, you'd watch him like a hawk. Do the same for the autopilot.

Be certain that you understand the effect of various failures on the autopilot. From which instruments does the autopilot derive information? And what does the autopilot do when an instrument fails, from loss of power source or any other reason?

One area of primary and almost necessary use for an autopilot is when operating IFR in marginal VFR weather conditions. Given, for example, virtually clear skies but limited visibility in summertime haze, the autopilot can be given the primary task of maintaining heading and altitude while the pilot keeps up an active scan for traffic. When flying in such conditions there is just no way to do a precise job of hand-flying the airplane while also doing the IFR chores and a thorough job of looking for traffic.

## FLIGHT DIRECTOR

The flight director is an autopilot adjunct that uses the information given to the autopilot to command the human pilot to move the controls as necessary to make the airplane fly as per the program. A virtual necessity in more sophisticated airplanes, flight directors make hand-flying any airplane much easier. Most come in a package with an autopilot: almost no airplanes have a flight director with no autopilot.

A flight director will tell of a straying from the desired condition before most pilots would catch on from using raw data, and it then tells how much attitude change is necessary to properly move the airplane to the selected condition. It is almost like having a flight instructor there, telling you when and how to move the controls.

## DME / RNAV

Distance-measuring and area-navigation equipment both add a great deal to an airplane's intelligence in IFR operations. They also make available more instrument approaches. Area-navigation gear can often be used for a straight-in approach where only a circling approach is available with a basic IFR package. They are things that you can do without, but they are also useful enough to be high on almost every IFR pilot's shopping list.

DME can be a very special IFR asset in a slower airplane. With it you can note any drastic change in groundspeed very quickly and make any appropriate changes in plan. I got a good example of this at the tail end of an IFR flight in a Skyhawk. Everything had been going well, and I had pegged the fuel remaining on landing at about the minimum one-hour reserve. The destination weather was good enough to consider it okay without an alternate. Then came a few minutes of wet and bumpy flying. Such can suggest that the airplane has passed through a front and that the wind might change. Indeed it did. The groundspeed dropped from 100 knots down to 75 knots. My hour's reserve was shot, a landing was dictated for fuel, and the DME had told me of the situation almost immediately. In a faster airplane, where any head wind is a lower percentage of cruising true airspeed, a DME is less important in this regard. But it can still save a day.

The DME also helps in picking the best IFR (or VFR) altitude. Just watch the groundspeed while climbing for an idea of what is going on. If you know what the true airspeed is in climb, and if a relatively constant true airspeed is maintained in the climb, you can get a good idea of the winds. It doesn't happen frequently, but there are times when the winds change substantially with altitude.

A final word about DME. It has been said that the DME is one of the least reliable and most expensive pieces of avionics gear to repair. That might be true, if for no reason other than the basic complexity of the DME. However, the contemporary DMEs that I have owned have given good service with reasonable maintenance requirements. When they get old the situation might change, but I put 1,200 hours on one without any trouble.

## BIG THINGS

De-icing equipment and weather radar have long been available for light twins and are just now coming on strong for singles. As an alternative to radar, a device that utilizes standard ADF antennas and plots lightning strokes on a cathode ray tube came out in 1977 and offered the pilot of the single-engine airplane a thunderstorm-information system without the antenna-installation problems of radar.

These items represent a large investment, and while many pilots mistakenly believe that they offer absolute answers to the ice or thunderstorm question, they do not. De-icing equipment is something to use while fleeing ice. The best operational procedure is the same with it or without it: Start trying to get the airplane out of the icing condition as soon as it is encountered. De-icing does make an icing situation less critical, and an airplane with a complete and tested set of gear (approved for flight into known icing conditions) can indeed slog along in most icing conditions for a long while. But the frozen stuff will still collect on unprotected surfaces and create drag. And there are icing conditions severe enough to best even an airplane with all the available equipment. I'll always remember seeing the remains of a de-iced twin that hit in the vicinity of the outer marker on an ILS one very icy night. The equipment was functioning and the engines were developing full power when the airplane settled to the ground—sheathed in ice except for the protected areas.

Radar, or any storm-information device, is perhaps even less effective than de-icing equipment in its application against an operational consideration. When a pilot gets such weather-information equipment for the first time, a very common comment is that situations are handled in the same manner with it as they were handled without it. No magic, in other words.

No electronic device can be used to safely penetrate thunderstorms in a light airplane. The meteorological fact of life is that thunderstorm avoidance is the only sane way to fly, whether you use electronic devices or not.

These things don't tell you how to penetrate or to avoid weather on any sort of absolute basis. All that's available is information on the location of precipitation in the case of radar, or information on the approximate location of electrical activity in the atmosphere in the case of the device that plots lightning strokes. This information must be combined with a visual assessment of the situation and a very strong knowledge of the nature of thunderstorms to develop an idea on what will probably be the smoothest path. That path might even take the form of a retreat to an airport away from the area, and recognizing that possibility is an important part of using the equipment. None of which casts any shadow on de-icing or weather-information equipment. I would like to have both in an IFR airplane.

## TOTAL KEY

The airplane and each bit of equipment added to it come to a total, and there will always be compromises. Put in too much gear and the airplane might cost a fortune and weigh so much empty that the payload and/or range is on the short side of the average mission. Then the question becomes whether the heaviest and the most expensive item makes a contribution in proportion to its weight and cost. If it appears that the item in question makes possible one trip a year, or would result in your being able to conduct instrument operations to a couple of extra airports a year, is that worth the financial and weight compromises? Might the money be better spent on an airplane with more range, better climb or ceiling, or a higher cruising speed?

After subjecting the heaviest and/or most expensive piece of gear to that test, go through it with some other items considered optional.

There are plenty of choices. For a practical result, recognize that instrument flying introduces some factors that are not strong in VFR flying, and choose the machine for the mission on a methodical basis.

# 14

# THE RISKS
# AND REWARDS

In aviation, as in so many other things, a personal assessment of the risks involved is based more on hearsay, guess, and assumption than it is based on fact. This is both unfortunate and dangerous. It is unfortunate because in flying with unjustified hangups we rob ourselves of utility, or at least we create unnecessary tension when doing certain things. It is dangerous because if we don't understand the risks, we can hardly be expected to guard against them in an intelligent manner. A lot of pilots run high risks while going through a rather continuous version of building the levee on the other side of town from the river. They placate fears that have no basis in fact while ignoring the things that are proven killers.

What do I mean? The best illustration relates to the number of engines on an airplane. How many times have you heard people suggest that a person is crazy to fly IFR in a single-engine airplane? I have even seen FAA employees write newspaper articles saying that single-engine IFR is not a wise thing to do. Such observations are based on ignorance of actual risks, not on the experience of real people flying real airplanes. The way to prove this is to examine the accidents involving IFR flights in a one-year period and learn where the general aviation pilot found the highest degree of actual risk in IFR flying. This examination is based on National Transportation Safety Board records, and includes the flights noted by NTSB as being on an IFR flight plan.

## IN 1975

There were 4,237 general aviation accidents in 1975, and 675 of these were fatal. Of the totals, 106 appear to have been on an IFR flight plan, with 60 of the IFR accidents involving fatalities.

There is a rather clear message there. Pilots on IFR flight plans tend not to have the minor accidents that plague general aviation pilots flying VFR. When an IFR mistake is made, the consequences are likely to be serious. We'll see why as we look at the nature of the IFR accidents.

## MECHANICAL

Having started the discussion with the suggestion that the number of engines on an airplane does not have any effect on risk, it is only proper to first examine the IFR accidents that were related to engine failures or mechanical problems. There were fourteen such mishaps, and five of them involved twin-engine airplanes. Five of the fourteen mechanically related accidents were fatal, and two of the fatal five were in twins. Two of the fourteen accidents involved injuries, and one of these was in a twin. The active IFR single-engine fleet outnumbers the twin fleet, so there is nothing in those numbers to suggest that there is any unusual risk found in single-engine IFR.

Closer examination of the accidents puts this in better perspective.

Of the three single-engine airplanes involved in fatal mechanically related or engine-failure IFR accidents, only one was an airplane that is in current production. It was a Bonanza, lost in a thunderstorm, and the sole mechanical relationship to the accident was an inoperative artificial horizon. It had no connection with the vacuum pump, which was apparently operating normally. Such a malfunction doesn't actually relate to single-engine—a Baron with an inoperative artificial horizon might not have fared any better. And indeed there is a strong chance that a Bonanza or Baron with an operative artificial horizon might not have made it through the storm, but NTSB listed the inoperative instrument as a factor, so the accident is considered here.

The next single was a Mooney M-22 Mustang, a pressurized airplane that was produced only for a short while. The problem was an in-flight fire, followed by an emergency descent into terrain. Would the outcome have been any different in a twin? If you had one engine out of two blazing away, wouldn't you likely head for the ground in haste?

The third and last single was a Comanche. The problem

was with carbon monoxide poisoning, a factor that would not discriminate between VFR and IFR. It is in fact a more likely event in a single than a twin, because of the exhaust-muff heaters in singles, but there are recorded cases of gasoline heaters in twins malfunctioning and causing carbon monoxide problems. So this can't be chalked up as something solely related to the single.

Note that not one of these fatal accidents actually involved the mechanical failure of an engine.

Neither of the twins involved in fatal mechanically related accidents was a current production airplane. One was a De Havilland Dove that had a complete electrical failure at night—something people worry about more in singles than twins—and flew into the ground. The other was a Beech Travel Air that had a partial power loss, perhaps from carburetor ice, and crashed during an attempted forced landing. That accident in a twin would more nearly follow the scenario that many would apply to the single.

The nonfatal accidents related to mechanical or engine troubles follow a pattern that you might expect. A couple of twins were wrecked because one engine didn't respond on a missed approach and the airplanes wound up in a ditch. Two singles were damaged in precautionary landings following battery or alternator problems. Two singles were landed off-airport following engine failures due to carburetor ice. One single was landed off-airport following an engine failure induced by icing of a fuel-tank vent. One pilot used all the fuel in a twin and had to land it in a randomly picked area at night. The mechanical relationship with that last one was in a failure of the avionics systems in the aircraft. There was also an avionics failure related to an accident in a single—the glideslope was not operating, and the pilot flew into the ground on the second attempt at a night ILS.

That covers the mechanics. The fatal accidents in both singles and twins were not really related to the number of engines on the airplanes, and the cases of engines failing in singles and leaving the pilot without propulsion were related to the pilots' failure to properly deal with icing. I have done studies of other years and the picture is always similar. Occasionally there might be a serious accident or two that is related to a mechanical failure of the engine in a single, but the same is true of the twin. That is not to say that some engine-failure-related risk does not exist. It surely does, single or twin, but the record suggests that the IFR risk in both types is both equal and minuscule.

## MONTHLY

The time of year has a lot more to do with IFR accidents than mechanical factors. January, February, and March accounted for almost half the accidents. That isn't much of a surprise, because the weather during those months is more conducive to IFR operations.

## JANUARY

Ice is a big factor in January, naturally, with 40 percent of the accidents so related. Many of the ice accidents are not serious, though, because they are related to hard landings with the airplane frosted. The main damage in such an event is to the landing gear and to the pilot's pride. There were three classic cases of the airplane's becoming iced to the point that it couldn't stay up, and there was one in which the pilot spun in while maneuvering to land with a load of ice. Even in January, a bigger deal than ice was the simple act of flying into the ground after descending below a safe IFR altitude. The fact that this is the number one problem even in the frostiest month of all clearly outlines an area of maximum risk.

The ice problem recedes after January and settles into the pattern it has followed for years, with most ice-related accidents occuring after the pilot gets near an airport and stalls the airplane while maneuvering for landing.

Before leaving the risks of ice, there is a final point to consider. We often think that the ice problem can be solved with money: Buy de-icing and the problem goes away. In 1975 there were twenty-five ice-related accidents (eleven of which were fatal), and eighteen of the twenty-five were in twin-engine airplanes, many of which carried de-icing equipment. It is true that a higher proportion of the icing accidents were serious in single-engine airplanes, but that factor might be influenced by things other than equipment. Some of the singles, for example, were flown by pilots with relatively little total experience. These people's response to the icing might not have been as prompt or as good as the response of a more experienced pilot.

The prime point is that having a twin, and even having de-icing equipment, does not constitute a purchase of safety. I'd lots rather fly away into potentially icy skies with a super-conservative pilot flying barefoot than with a de-icer boot-equipped daredevil.

―――――――― *THUNDERSTORMS* ――――――――

It is quite logical that the thunderstorm risk peaks in the summertime. The worst month in 1975 was July, and all but two of the eleven IFR thunderstorm-related accidents occurred in May, July, and August. There were none in June, and, in fact, I didn't find any IFR accidents in the month of June.

Why no IFR accidents in June? Most likely it was just a fluke of the year. The total number of accidents is small enough for none to occur in a month when the weather is reasonably good, and perhaps the benign nature of the lovely month of June, sandwiched between the last violence of springtime storm systems and summertime thundering and snorting, made possible no mishaps in the month. It would be nice, though, to speculate that if any one month can be IFR accident-free, any other month could also be accident-free.

Nine of the eleven thunderstorm accidents were fatal. The two that involved no injury might be considered unusual. In one, a light twin hit wires and poles while attempting a landing on a drag strip at night in a thunderstorm area. In the other, a Twin Comanche suffered airframe damage but was landed safely.

Five of the nine fatal thunderstorm-related IFR accidents

in 1975 involved Model 35 Bonanzas, and in every case the airframe failed in flight. Ironically, the Bonanza is licensed in the FAA's utility category, meaning that the airplane has a higher limit-load factor, which reflects greater airframe strength.

The nature of the Bonanza accidents suggests that in every case the pilot had lost control before the failure. One of these was mentioned earlier when discussing the malfunction of its artificial horizon. In the others, no factor was mentioned that might have contributed to the pilots' loss of control.

The pilots flying the Bonanzas represented a wide variety of experience. One had no instrument rating even though an IFR flight was being conducted. One was a 3,000-hour commercial pilot with a lot of Bonanza experience. Two were pilots with relatively low (under 400 hours) total flying experience. One was an experienced private pilot.

When the other four fatal IFR thunderstorm accidents are considered, they add something to the picture. Like the Bonanza, the other four airplanes were retractable-gear aircraft. A Mooney, an Aerostar 600, a Beech 18, and a Piper PA-23 were involved.

The fact that all airplanes lost in thunderstorms in this year were retractables strongly suggests that the risk involved in flight in the vicinity of storms increases dramatically with aerodynamic cleanliness. The Bonanza involvement suggests that there is no salvation in simple airframe strength. Once an airplane is out of control, the rapid speed buildup encountered in a clean airplane will quickly negate any strength advantage. There's never been an airplane built that couldn't be destroyed given the right (or wrong) set of circumstances, and if you fly one that is tough enough to hang together in a thunderstorm-induced loss of control, the outcome might be different only in the airplane's breaking as it hits the ground instead of before it hits the ground.

The record emphasizes a message for pilots of retractables in general and Bonanzas in particular. If the level of turbulence is enough to make you the least suspicious of a lateral upset, fly with the landing gear extended to avoid a rapid speed buildup in case of a loss of control. Back in the good old days people often extended the gear in turbulence because they thought the airplane might be more stable in that configuration, but that's not the reason it is a good idea. It's to give you maximum advantage if you drop the ball. Another item: In reports on storm-related accidents there is often a witness statement to the effect that engine sounds were loud, as if full power were being developed. What likely happens is that the pilot just leaves power where it was when the upset commenced. A better reflex action would be to close the throttle in reaction to a rapid speed buildup in an upset. Leaving power on will only aggravate the airspeed buildup. We have such good power response in light piston-engine airplanes that there is no bad byproduct (other than rapid engine cooling) from closing the throttle when it appears necessary, and then coming back with power as required once the situation has improved.

Don't be tempted to use flaps in case of an upset. Extension of flaps lowers the limit load factor (the number of G's the airplane will withstand) on most airplanes, and you are likely to need all available airframe strength to make it through the recovery.

To have a look at speed buildup, for just a moment put the nose of your airplane down about 10 degrees at normal cruise, clean configuration, and note the rate at which the airspeed increases. Then try the same thing with the gear down and the power off. Big difference in the behavior of the airspeed indicator.

As a final thing in your favor for a turbulence-induced upset, take a little acrobatic training. This will teach you about the way the airspeed increases when the nose is well down, and how to roll an airplane back upright from an inverted position. If you can, rolling one upright is by far a better deal than a split-S when upset to an inverted position, but apparently most losses of control in severe turbulence progress so rapidly that the airplane is in both lateral and longitudinal trouble very quickly. Then the drill is to manage airspeed with all available means, roll the wings level, and pull out of the dive as gently as possible.

There's turbulence around the mountains that is not thunderstorm-related, and this led to two airframe failure IFR accidents in 1975. One was a Seneca and the other was a Comanche.

With thunderstorm, ice, and mechanical risks considered, we've accounted for fifty airplanes that bit the dust in 1975. Of those, more than half (twenty-eight) were twins. The pilots of single-engine airplanes are doing pretty well at managing IFR risks so far.

---

### ROUTINE

With those things behind us, IFR accidents take on a predictable sameness that is very discouraging. Time after time we find that the pilot simply flew the brightly painted airplane into the whole world. No contest. No contest at all.

An Aztec hit a fifty-foot tree on the third attempt at a night VOR approach in below-minimum weather conditions.

A Cessna 210 descended below the decision height and hit terrain.

A Mooney descended below the minimum descent altitude and hit trees two miles short of the runway at night.

An Aztec flew into a lake on a night approach.

An Aero Commander hit a tree when below the minimum descent altitude on a night approach.

A Travel Air crashed three miles from the airport on an ILS approach at night—on the localizer but quite below the glideslope.

A Mooney missed the approach and then flew into terrain on an improperly executed missed approach at night.

One night a Dove descended below the minimum descent altitude and hit wires a mile and a half from the end of the runway.

A DC-3 hit a tower two miles from the runway. The weather was below minimums for the night approach.

A Cherokee hit a mile short of the runway on a night ILS approach in minimum conditions.

A Cessna 172 went below minimum descent altitude and hit a mile and a half from the runway at night.

A Beech Sport hit trees three miles from the airport on a localizer approach at night.

A Skymaster crashed after letting down at an airport where there was no published approach. You guessed it: at night.

A King Air went below minimum descent altitude and hit trees a mile and a half from the end of the runway at night.

An Alon A-2 hit trees while attempting a zero-zero approach at night.

A Merlin flew into the ground a half mile from the runway on a night circling approach.

A Baron descended from the assigned altitude and flew into terrain for undetermined reasons.

A Duke flew into the ground two miles short of the runway while attempting an approach in below-minimum conditions at night.

A Piper Arrow flew into the ground short of the runway while attempting an approach in below-minimum conditions at night.

A Cessna 320 pilot flew into the ground on the third try at an ILS approach.

An Aztec pilot descended below the minimum descent altitude and flew into the ground on a night approach.

A Baron descended below minimums, hit the glideslope shack, and crashed.

A Beech Sierra flew into a hill while below minimum descent altitude on approach.

A Bonanza flew into a mountain after the pilot became lost on an IFR flight.

A Bonanza pilot descended into water on final.

A Cessna 210 pilot crashed on ILS final approach course. This was the only IFR accident of the year in which alcohol was listed as a factor.

A turboprop Rockwell Commander descended into terrain while flying a homemade instrument approach.

A Rockwell Commander 112 hit a mountain while descending at night.

A Bonanza descended into the ground on the second try at a below-minimums ILS approach.

A King Air pilot descended below the minimum altitude for the segment being flown and hit a radio tower.

A Cessna 402 pilot descended below the minimum en route and hit terrain.

A Bonanza pilot flew into the ground on a night circling approach.

Compare the length of that list with the brevity of the list on mechanical problems, ice, and thunderstorms. It does rather make a mockery of the things that many of us buy and do allegedly to minimize the risks in IFR flying. It also illustrates a great lack of discipline among general aviation pilots.

We shoot more instrument approaches annually than air-carrier pilots, but not a lot more. Very occasionally—every year or three—an air-carrier crew might fly into the ground. General aviation pilots do it with a grinding regularity. Even the professional general aviation crews do it with an alarming frequency. Note that there were five turbine-powered business aircraft on the list of premature arrivals.

It might be suggested that general aviation pilots face more inherent risks on approach than airline pilots, because more of our approaches are of the non-precision variety, without vertical guidance. Also, we often shoot approaches to airports without weather-reporting services. That is an invalid theory, though. It is never necessary to allow these things to affect the safety of a flight. A non-precision approach to an airport without weather reporting offers no inherent risk. If flown correctly, the airplane will remain at a safe altitude until the pilot has the runway or its light in sight. Cut and dried. The non-precision approach will not allow operations in conditions as low as a lot of pilots would like, though, and therein lies the risk.

The risk is self-imposed and easily understood (and avoided). A pilot's violation of procedure or good operating practice is a necessary precursor to flying an airplane into the ground. The record clearly shows, too, that the visual tricks of darkness apparently tempt pilots into that dark never-never land that is populated by trees and other obstructions. Minimizing the approach risk is a matter of proficiency and an absolute level of discipline.

## ——— THE MISCELLANEOUS CAUSES ———

The items covered to this point account for the great majority of IFR accidents. Only one other cause follows a pattern, and it is directly related to pilot proficiency or technique.

Spatial disorientation with no contributing factor such as turbulence or a mechanical failure is listed as the probable cause of a number of accidents. The pilot simply did not pay attention to flying, and the airplane went off in a direction of its own choosing. No doubt the pilot recognized the problem at some point, but was either unable to recover before the ground intervened or was unable to recover without breaking the airplane. It is interesting that the number is so high. Almost as many general aviation IFR pilots—eight—lost control and crashed on general principles in 1975 as lost control and crashed because of involvement with thunderstorm activity.

The list of spatial disorientation accidents teaches its lesson well. Five of the eight accidents occurred at night. There is no question that night is a more opportune time for disorientation than day. The chores that take attention away from the instruments, such as chart reading, are more difficult and more likely to hold a pilot's attention at night, too.

Half the accidents involved fixed-gear airplanes, which means that an airplane does not *have* to be aerodynamically clean to get away from you if left untended.

Two of the pilots were not up to snuff. One was without an instrument rating even though he was on an IFR flight plan; the other was not current.

The other pilots were experienced, ranging from 500 to 16,000 hours total time. The latter is just one more illustration that the hours logged hardly matter as much as the hour being flown.

───────────────── *NO REASON* ─────────────────

There are always a few cases in which IFR airplanes fly into the ground for no apparent reason. They just hit level, going fast. These usually happen at night, and there is often some suggestion of pilot fatigue in the findings. The message is to remain on the ground when tired, regardless of the type of flight—VFR airplanes also fly into the ground—but if a pilot must fly on, I'd suggest that possible fatigue is just one more reason to always operate IFR at night. The demands of IFR can help keep a person more alert, and the requirement for communications can help keep a pilot from falling asleep at the wheel. Certainly we should never think of an accident as having been caused by a pilot flying *IFR* when tired. It would better be thought of as just plain flying when tired.

An occasional IFR flight winds up in the ditch at the far end of the runway after a high and fast IFR approach. These tend to be minor, and occur less frequently than you might imagine. There do seem to be a number of missed-approach accidents in twins in which a recalcitrant engine is listed as a cause. The accidents usually involve nothing more than a slow-speed trip into the boondocks, and no cause is generally established for the engine problems. One might only speculate that there was really nothing wrong with an engine, and that the cause of the accident was procrastination and a late decision to go around after a poorly planned approach.

There were three cases of IFR fuel exhaustion in 1975, and one case of fuel-system mismanagement causing an accident.

A number of relatively minor accidents don't really relate to instrument flying but were included in the total because the airplanes were on IFR flight plans. Snowbanks are a rather frequent factor in wintertime operations, and there are a few instances of an airplane not getting off the ground to begin an IFR flight because of ice, snow, or frost on the wings.

## TALLY

Having examined the bulk of the accidents, it is possible to examine the true risks involved in IFR flying.

The myth that single-engine IFR carries with it some unique risk is not supported by fact.

The myth that thunderstorms gobble up general aviation airplanes at a great rate is modified. Nine were lost to storms in 1975. That's bad if you happen to be one of the nine, but it is a relatively small part of the total picture. This in no way minimizes the hazard of the beasts, but it does put the risk in perspective and gives some comfort to the IFR pilot who studiously attempts to avoid storms and who puts great effort into maintaining proficiency at flying in turbulence.

Ice is a problem of about the same magnitude as thunderstorms.

The primary risk is in the approach phase of flight, and is best managed by simple discipline.

## "HARD" IFR?

In leaving the subject of risks, I would like to cast a harpoon at some terminology that has been floating around general aviation for years. Some say that they would not fly "hard" IFR or that they attach certain conditions to the machinery used to fly "hard" IFR.

This is fiction that should almost provoke laughter. All actual instrument flying is created equal. It is quite impossible to put a grade on an upcoming flight, because of the basic limitations of information we receive in a preflight briefing. What might appear from the ground as "hard" weather could actually turn out to be a marshmallow. And what could appear easy from a distance can turn out to be wet and bumpy, or icy, when reached. The forecast 800-foot ceiling might indeed be 200 when you get there.

The situations and conditions that are found in instrument flying are dynamic, and they defy advance classification. When a pilot sets out to fly instruments, the risks are best managed by understanding, by flying with an open and disciplined mind, and by maintaining the ability to meet every challenge.

## THE REWARD

How reliable does the light airplane become when operated IFR? What is the measure of the reward? Well, it is not possible to put a set percentage on a higher degree of reliability, because a lot of factors can cause the figure to vary. Trying to operate IFR in the Rocky Mountains area in an airplane without turbocharging isn't something to bet on, for example. Too, some pilots are more reluctant than others. Some IFR operations lend themselves to cancellation because of airport location or configuration. Certainly you are going to miss approaches at an airport with a non-precision approach much more often than they might be missed at an airport with a full ILS. It is possible, however, to offer experience as a guideline to the reward that might be found.

My IFR flying started in 1955, and since that time I've completed a very high percentage of the planned trips on the prescribed day. Home base for about half the time I've flown IFR has been in the New York area; my base was in Arkansas for the other half of the time.

In reviewing a year's worth of activity in the northeast, flying a Cessna 182 from an airport with a non-precision approach, I found that 406 cross-country legs were planned during that year. Of these, 384 were completed with little or no delay—meaning that I got where I was going at about the time I thought I would get there. Twenty flights were delayed substantially—I got where I was going on the day that I planned to get there, but I was very late. Two flights were scratched during the year. The number of outright cancellations in this year was a little below average. Looking at a lot of years, I find an average of four trips a year that are canceled because of weather.

The reasons for cancellation are what you might expect. Looking at cancellations from four years, I found fifteen. I was based in Arkansas during this period, and thunderstorms were the primary cause of scratched flights. Eight of the fifteen were prompted by either squall-line or widespread imbedded-thunderstorm situations. A couple of snowstorms are on the list, along with three below-minimum situations and an episode with a rough engine. High surface winds prompted one cancellation. (The winds also caused postponement of the meeting I was going to, so does that one really count?)

Almost all my trips are planned at least a week ahead of time, so you can't say that I must just plan trips when the weather is good. That's just not so.

In basing at an airport without an ILS, I have had numerous instances of landing at a nearby airport with an ILS, but I don't count this as a cancellation because I get where I am going (home) as scheduled with only the minor inconvenience of having to hitch a ride to pick up my car.

Cancelling 1 or 2 percent and delaying 5 to 8 percent of

the flights has, in my opinion, been quite adequate for avoiding real misadventures. What would the cancellation rate have been in flying solely VFR? I'd guess as high as 20 percent, so the reward is quite great.

## ———— HONESTY IS THE POLICY ————

A pilot must be inwardly honest when making the decision to undertake or continue a flight. And a good grade on the decision comes from how you feel about the flight while en route, or after landing. If there is any feeling that the things are (or were) chancy, that the completion was due even in small part to good fortune, or if you were nervous or ill at ease over the proceedings at any time, then the reward sought was excessive. Reaching the destination was simply not worth it.

I don't mean by this that you can't be conducting a good safe operation and still have moments of super-alertness when flying IFR. The greatest challenge in flying is to deliver the goods. The destination is straight ahead. Go that way, press on. Be challenged. Be alert. But also be ready to accept the evidence as conclusive when it stacks against going on to the destination. Part of such evidence is in what you see and hear; part of it is in how you feel.

A friend of mind told me of a flight that he made that proved something about this to him. He was flying a light twin and was pressing hard to get home after being on the road for a week. The carrot was the warmth of family and home. The stick was a great collection of thunderstorms over the Appalachians.

He kept on going toward the area of weather. The air traffic controller told him of the weather, and my friend asked for a vector to the best-looking area.

Mistake number one was on mental attitude. There is quite a difference between the "best-looking" or "lightest" area of weather return and a good place to fly. When fooling around with weather, these relative terms are often not acceptable. Had my friend asked if there was a *good* area to go through, the answer might well have been "No."

When he later related the flight, I got a strong feeling that he was apprehensive as he got close to the weather. Still, the potential reward outweighed the risk in his mind, and he plunged on.

Then came the first cell. He said that the airplane was ascending even with the throttles back. The rain was torrential, and the turbulence almost unbelievable to him. The thought that this was a very hazardous situation surely crossed his mind, as did a wish that he had not proceeded.

The pilot hung on, though, and emerged from the other side of the cell unharmed but well shaken. The next question to the controller was about any more cells ahead. There were a lot, and the pilot's mental conditioning was now such that the potential reward of getting home soon wasn't worth the risk. That sample was convincing, like being clubbed, and he sought and found a storm-free path to an alternate airport and went there and landed. He had stepped across the line, and he knew it.

## THE AIRLINES

The air-carrier system does a magnificent job in the risk-and-reward business as it relates to IFR in general and thunderstorms in particular. The risk in their flying is very, very low, and the reward is very, very high. They seldom cancel for weather, but there is still a gray area in their thunderstorm operations, and it relates to all flying.

Thunderstorms don't often affect carrier operations, but occasionally they lose an airplane to a thunderstorm. And if they lose one every three or four years, would the reduction in risk found in following absolute thunderstorm-avoidance precepts be worth the reduction in the reward of schedule reliability? If indeed the air carriers stayed five miles away from all observed thunderstorm cells and twenty miles away when severe weather is forecast, some slight element of risk would be removed, but there would be many days a year of schedule chaos at the major airports of our land.

Too, it must be considered that the elements are dynamic. Pilots and meteorologists can only make their best estimate as to the severity of a bit of weather and base the decisions on that estimate. In cases where air-carrier airplanes are lost in thunderstorm areas, the pilots certainly did not think that the existing weather conditions were too severe to fly through, and in terminal-area accidents other airplanes ahead of and behind one struck down by a storm often proceed unhindered.

If you think I'm suggesting that there is a slight amount of mystery found in seeking reward while dealing with risks posed by the elements, you are precisely correct. Nobody will ever know all there is to know about airplanes versus weather, and if we are going to fly when conditions are inclement we'll occasionally lose one—and that even includes the best-trained air crews flying with the best equipment. It is often impossible to explain this to a layman. People who don't fly often believe that something like a thunderstorm is a given force that can be fenced, isolated, and avoided.

## BEST REWARD

IFR flying's best reward comes in benign weather situations. Here it is possible to plan a flight at a desired altitude and depart with knowledge that the flight can be flown at that altitude. That is in marked contrast to VFR flying, where available altitudes are often restricted by clouds.

IFR also offers a personal challenge and satisfaction that can only be considered a reward. To plan and conduct an IFR flight with precision is challenging and stimulating. From the first thought of weather when preparing for the flight to the personal debriefing that should be conducted after every flight, the operation is demanding. Flying VFR, you reach an accommodation with gravity. Flying IFR, the accommodation is with gravity, the elements, and the system. The total reward is much greater utility in a flying operation that involves very little inherent risk, especially when compared with scud-running in marginal VFR conditions. Once you learn to do it, it becomes an enjoyable exercise as well.

# 15

# KEEPING
# IT TOGETHER

Once a pilot gets an instrument rating, has available an airplane fit for the envisioned mission, and understands the risks, the task becomes one of keeping the act straight. The first and prime requirement for this is avoiding self-deception.

First, consider that there are days when some pilots should eschew instrument flying, and there are even some pilots who should abstain completely. IFR is both more demanding of perfection and less forgiving of error than VFR flying. If a pilot feels that instrument flying has had a best shot, but that performance during IFR flights is in doubt from start to finish, then keeping body and soul together becomes a matter of sticking to VFR until seeking more education. If that doesn't help, quit the IFR.

There's nothing to be ashamed of in acknowledging an

IFR weakness. Instrument flying requires a rather unique blend of skills and disciplines that is so foreign to some people's psyche that there's just no way for them to be comfortable with the activity. The real ace is the relatively rare pilot who is smart enough to stay away from instrument flying after recognizing that he or she isn't adaptable.

For the great majority, the task is only one of learning about the activity and maintaining a good level of proficiency.

Past the initial basic training, the best way to learn is by practicing, by doing, and by being critical of each performance. The latter is both effective and especially important. Any pilot can grade his or her own technique at instrument flying; anyone who can't is not really capable of the basic task.

## PRACTICE

Periodic practice is very necessary for every instrument pilot. It takes two forms. One involves the routine flying of an instrument flight. This can be skipped if a pilot is flying actual weather regularly. The other form involves the practice of things that are not routine. This is necessary for all instrument pilots.

If practice of the routine is necessary, it should simulate actual operations as nearly as possible. If the mission is two hours' hood practice, a flight plan should be filed and the flight should be conducted in the IFR system. The required safety pilot can meet the minimum requirements, but it is productive to take an instrument instructor along on occasional proficiency flights.

In two hours of practice it is easily possible to make a short cross-country and shoot several of the variations on instrument approaches. A full ILS, VOR, and NDB approach should be practiced. At least one circling approach is a good idea, and it is good to land out of some approaches and execute a missed approach in others.

If a pilot maintains a basic level of currency, this type of practice flight can be made in actual instrument conditions when the opportunity presents itself. Go to the airport early on an IFR Sunday morning and work around the airports in the area, actual IFR. It saves money—you don't have to have a safety pilot—and it is a much more valuable experience than hood time. And there is no question that staying current on approaches by shooting actual approaches is by far the best way to do it.

## THE GRADE

When examining the level of skill displayed during a routine flight, practice or otherwise, be especially critical. Routine flights are the ones during which everything works in your favor. If an ideal situation can't be handled with precise aplomb, look out when the weather or the machine starts to show a seamy side. Start at the beginning, with the preflight, and examine every action for any weakness. Remember never to leave with the thought that "it was okay because I made it." The alternative to making it is usually an infinite period of silence, so you'd better always "make it" by a very wide margin.

In studying accident reports, I often look at the pictures of shattered airplanes and try to imagine what was going on in the cockpit just a few seconds before the impact. In most IFR accidents, there was probably no apprehension. The pilot felt he was doing the correct thing. Trouble was, he hadn't maintained a critical curiosity about the situation and double- and triple-checked everything. Or the pilot was not following the rule to the letter. When your grades aren't straight A's, probe for weaknesses.

───────────── **NOT ROUTINE** ─────────────

I recall a routine flight that suddenly varied from the routine and gave me an interesting exercise in self-grading. At such times we need to be exceptionally critical of our performance and examine every aspect of the event.

I was flying a brand-new turbocharged airplane. The IFR clearance was to 14,000 feet, and during climb I was deviating around some rainshowers. I zigged when I should have zagged for one shower and wound up in some heavy rain and moderate turbulence. The rain started affecting the aircraft's static system. The airspeed, vertical speed, and altimeter were all showing signs of being bugged by water, so I selected the alternate static source.

Immediately after I selected the alternate source, the airspeed started increasing. Thinking that I was in an updraft, I reduced power a little. The airspeed kept increasing, and I became suspicious. It was well up into the yellow, yet there was no increase in sound level, the wings were level, and the airplane was in a normal attitude. There was also no sensation of an updraft. I know you can't fly by the seat of the pants, but there are certain feelings related to certain things, and it is hard not to be physically conscious of a strong updraft. The airspeed indication just had to be in error, but one doesn't take lightly any decision that an instrument is in error.

It was far too late for an A in the course when I related the airspeed problem with the alternate static source and returned to the primary source. This brought things back more nearly to normal. My grade dropped even more when I realized that I had subconsciously and continuously reduced the power as the airspeed increased and had slowed to a rather low actual airspeed in responding to the problem. On later investigation I found that the alternate static source in this airplane had been hooked up incorrectly and would cause erroneous indications on the pressure instruments every time it was selected.

The general lesson was that when something goes haywire, go immediately to the last thing that was moved or changed and see if that isn't what caused the problem. The specific lesson was that I had done poorly in using curiosity to quickly isolate a glitch. The example is illustrative of many situations in which we become mentally paralyzed. Watch for it in grading yourself. When it happens, a good scolding about the necessity for keeping the mind active is quite in order.

## THE DETAIL WORK

Back to practice of the routine: Demand perfection in holding altitude and heading, and in maintaining a proper airspeed. That is the basis of instrument flying, and if it is not done well the rest of the operation will probably follow a rather shabby pattern. Don't be satisfied with anything less than the proper position for navigation needle, either. There is a proper reading for each gauge for each phase of flight. The needles have to be somewhere, and they might as well be in the proper place.

Nobody is perfect, and all pilots wander and stray from the straight-and-narrow. In grading your own performance, consider how often you stray, how long it takes you to catch the problem, and how quickly and smoothly you return the needles to the proper position.

## THE HARDER PART

In practicing take advantage of the opportunity to experience less-than-ideal conditions. Something like the turbulence on a hot and windy day is too good to pass up. Shoot approaches. If you can do a good job of instrument flying in uncomfortable conditions like that, turbulence on actual IFR flights is likely to bring less of a tightening of the gut. The winter winds blowing over rough terrain also offer good opportunity for practicing instrument flying in turbulence. Imagine you are in the grip of a thunderstorm, relax, keep the wings level, and make peace with the machine.

Partial-panel practice is also essential, for it is the backup to the vacuum pump in a single-engine airplane. In a real situation this would be considered a semi-emergency. In other words, I'd accept special favors from a controller. But when practicing partial-panel, do it as if this is the way it is always done. Accept nothing less on partial-panel than the ability to shoot an ILS approach to minimums. Accepting anything short of that dramatically increases the risk created by vacuum failure.

Steep turns are good practice, for they teach discipline. Most of us don't do very well at 45-degree banked turns in clear weather; it is even more difficult IFR, because rolling into and out of turns can be spatially disorienting.

Unusual attitudes are also good practice, but don't let just any old safety pilot put you into any unusual attitudes. Reserve the practice of these for flights with an instrument instructor.

## ENGINE OUT

In a single, practice an engine-out glide. In the unlikely event the engine should fail, you might as well know what it is like to fly IFR in a gliding airplane. Practice the procedures for maximum glide distance.

Practicing emergency procedures in a twin is rather up to the individual. My personal feeling is that I do not want to fly IFR in a twin unless I feel I have the capability of shooting an ILS to minimums in the airplane with one engine feathered. The only way to know that is to practice. By so doing, some risk is accepted in the training environment to minimize risk in the actual environment. I am willing to do that as a favor to people who ride in airplanes with me.

As you work into and through a practice session, a good feeling comes as mistakes are made, analyzed, and corrected. It's almost like going to the doctor for a checkup. Put it all on the line. Hide nothing. Examine all the nooks and crannies. Don't do this just to satisfy the minimum requirements, either. Stay way ahead of the rules. Just as minimum altitudes represent the very lowest acceptable height above the ground, minimum requirements specify the lowest acceptable level of proficiency activity to satisfy the letter of the law.

Never fail to critique an instrument flight, practice or actual. And examine every part of the flight. If, for example, you are flying into a busy terminal and things seem not to go right, probe your mind after the flight for reasons.

## TOTAL

In the final analysis, instrument flying is a large part of the total when the transportation capability of pilot and airplane are considered. It has been said that a pilot without instrument capability is only half a pilot. That is very true if the purpose of flying is to get somewhere on a reasonable schedule. In the history of general aviation, IFR flying by nonprofessional pilots started almost exactly when the airplane started being considered something other than a toy. Since then, everything has worked for the better. Avionics equipment has become more capable at lower relative costs, airplanes have become more reliable, facilities have become very much better, and, above all, the user has been getting more value from flying.

# EPILOGUE

Some of the principles of instrument flying are worth reviewing over and over again. I list some of these here. The list is not offered as a complete one; rather, it is something for you to contemplate and expand to suit your need.

- Don't overcomplicate instrument flying. While it is different and challenging, and is not for everyone, it shouldn't be considered exceptionally difficult.

- Recognize that the basics are the cornerstone of IFR flying. Just as you can't fly at all if you can't take off and land, you can't fly IFR if you can't maintain the desired heading, airspeed and altitude, or rate of climb or descent.

- After the basics are mastered, strive for excellence in the application of these basics to navigation and aircraft handling during difficult times.

- Be wary of absolute pronouncements about the affect of

controls, such as: "The elevators control altitude." In truth, moving one control always affects more than one instrument. If you must think in terms of what controls what, think of the pilot as controlling the whole airplane. There are times when a pilot might forget that simple fact, but the airplane always remembers.

• Don't forget the importance of always knowing your position in space, especially in relation to the terrain. Immediately preceding an IFR accident, the answer to the simple question "Is this a safe altitude for this position?" is usually "No." Continually ask that question about altitude. If the answer is not an unqualified "Yes," better get with making it so.

• Remember that IFR is a precise business. Numbers on charts mean exactly what they say. Minimums are the absolutely lowest safe altitudes at which you may fly without visual reference. When flying an instrument approach, be aware that this has proven to be the most lethal phase of IFR flight. *Never ever* leave the minimum descent altitude or the decision height unless the runway is in sight and you are in a position to make a normal landing. The rules aren't that strict, but do yourself a favor and settle for nothing less than the runway itself.

• Recognize that the required meteorological knowledge for issuance of an instrument rating is sketchy, even though it might not seem so at first. The IFR pilot must continually study weather, and any time you feel that an FSS briefer knows more about meteorology than you do, it is time for you to really hit the books. It is your responsibility.

• Beware the biggies, ice and thunderstorms. Study, understand, be curious, and *never* fail to respect their destructive powers. Never procrastinate when dealing with either.

Remember that avoidance is the only policy, but do not give up if avoidance efforts fail and you wind up in thunderstorm or icing situations. Failure is not automatic unless the pilot accepts it as such.

• Options are a big thing in IFR flying. Never intentionally operate without an option. The key is in always being able to say: "If Plan A doesn't work, I can revert to Plan B."

# INDEX